electronic harassment 89
fear of one's power 135

Playing GOD

Biological and Spiritual Effects of Electromagnetic Radiation

A Journey of Discovery

Disclaimer – *Playing GOD* provides a health and spirituality perspective with emphasis on the author's direct experience. This book is for informational purposes and not provided to diagnose, treat, prevent or cure any illness or disease. Dharam House and the author make no claims as to the accuracy or suitability of any described course of action. Everyone has a different 'prakriti' or constitution and therefore their own unique journey with illness and recovery. Your intuition and knowing are invaluable as is on occasion a medical doctor's opinion. Permissions have been sought for reproduced figures or text otherwise please contact the author.

Playing GOD

Biological and Spiritual Effects of Electromagnetic Radiation

A Journey of Discovery

Benjamin Nowland

Cover concept by Benjamin Nowland
Book design by Prue Mitchell pm3artdesign.com.au
Author photo by Kate Nutt

ISBN: 978-1-925341-24-9
Published by Vivid Publishing
P.O. Box 948, Fremantle
Western Australia 6959
www.vividpublishing.com.au

Cataloguing-in-Publication data is available from the National Library of Australia.

Acknowledgements

Thank you, Nature, for guidance every moment I am humble enough to listen. Thank you to the retreat for being a healing sanctuary after my radiation pollution sickness decline. Thank you to Neil Cherry, Les Dalton and David Hollway for their past work on the topic. Thank you to authors and teachers pushing the boundaries of perception. Thank you to all co-creators of health and harmonious development during this accelerated period.

Contents

Introduction

GRANDMOTHER KNEW BEST

ONE OF MY GRANDFATHERS WAS AN ILLUSTRIOUS fisherman and the other a master crabber. By the time I was six I was feeding the family and selling surplus door to door to bring in some cash. The grandmothers were a little less active but to me were the wise ones in the family. They each had a veggie patch and the sea breeze gently wafted a scent of citrus onto their porches. They dolloped butter on scones, toast and porridge. There'd be a knitted wool beanie for Christmas and Sunday lamb roasts were finished with ice cream and jelly. They mandated we sit well back from the television. The televisions of the 1980s and into the 90s were cathode ray – high emitters of low frequency magnetic field radiation. They suggested we play outside and we needed little nudging. In those days there were no telco towers beaming microwaves into our backyard. There were no Wi-Fi emissions from every household and café. Parks and school ovals were health enhancing zones with electromagnetic field levels close to the background level of Nature. These areas free of electromagnetic pollution I call Zero EMF Sanctuaries.

Then came the pro-margarine propaganda. As we sauntered down the cold foods aisle we were confused. Do we listen to Grandmother and stick with butter or listen to the men in white coats on the tele? Skewed science smothered us in margarine for two decades in order to sell a product. The margarine propaganda machine then slowed and again butter was better. Grandmother did not touch margarine in those two decades. She had a direct knowing of what was healthy and what was damaging. When we align with our intuition we all have this direct knowledge of what is healthy and what causes us harm.

Technology and societal 'noise' means it is not always straightforward to connect to this knowing. The noise includes so-called experts telling us we need more technology and microwave exposure. Apple has surpassed Exxon Mobil as the world's largest company. They recently took the record for the largest profits ever at US$18 billion for three months work.[1]

Meanwhile a growing contingent of society is disinclined to be swept up in mobile-mania. Some of us have tried margarine and gone back to butter.

REWIRING

When I started writing **Playing GOD** my focus was on rewiring the external. After a mid-2013 move to inner city Sydney I experienced acute symptoms of radiation pollution sickness over a period of four months. The only place I felt better was away from towers, mobile devices and Wi-Fi. My symptoms corresponded with scientific evidence for biological effects. Headaches, insomnia and brain fog cleared when I moved out of the city and into a Zero EMF Sanctuary rural retreat. The symptoms eased but I was now a functional hermit. Cities, office work and flying were out of the question. Radiation detoxing reduced the intensity of biological effects that took me to the edge in Sydney but my mental health symptoms continued. I'd not experienced sustained depression like that. Every time I reached a bottom I found myself knee-deep in mud and sinking even deeper. It was apparent that detoxifying from electromagnetic radiation (EMR) pollution was a crucial part of my journey but further discoveries were required. My system still held stress and trauma.

I wanted to be able to spend an hour in the library or hop on a plane to visit friends and relatives. Watching a movie at the cinema was no longer an option. Mobile phones on silent continue to emit radiation. Living as a radiation refugee, away from electromagnetic sources, my health was a five out of ten. I didn't want to live a life at five out of ten. What was the missing piece? During the ensuing year I searched and spent thousands of dollars on slickly marketed 'solutions'. Every product failed. I consumed litres of snake oil for the desperate. Anxious for a solution I flew to Bali seeking a 'magic pill' restoration from the revered energy healers. Temporary relief was obtained but symptoms soon returned as I motor-biked back to my accommodation. I returned home to Australia frazzled after the high radiation flight and with only one option left.

An internal rewiring was required. The missing piece was inner peace. Decades of seeking and learning came together. I refined techniques to release long-held stress and trauma which had surfaced for collapse. Transformation was the side effect of a deep internal journey while acting on the external challenges to the extent I could. Today I choose to spend far less time in high-exposure environments than I once did. I choose not to live next door to a mobile tower. After five months mobile device free, I now choose to own a mobile that is switched on an average of five minutes a day.

ABOUT THIS BOOK

Electromagnetic radiation is invisible and this sometimes scares us off from attempting to understand it. 'Experts' and industry tell us we cannot understand it and feed us at best disempowering information. I wrote **Playing GOD** to empower a general reader by simplifying electromagnetic radiation to make it accessible. I sought to practically connect electromagnetic radiation to our lives. This book is the culmination of two years of direct experience, observations and correlations. Part of that was the Sydney EMF Experiment – after symptoms commenced on moving into an inner city apartment I turned my decline into an experiment. I documented relationships between microwave radiation and my health and I committed to mastery of the subject. After a period of four months I had no option but to leave Sydney for the countryside. A few weeks later I had recovered sufficiently to begin writing but it took another year and a half to regain thriving physical and mental health. This came after a comprehensive search for answers and my own discoveries, observations and surrender.

My first-hand experience is supported by sound research with the foundation of an honours engineering degree and years of work as a multi-modality healing practitioner. Some of the experiences in the book are of my clients.

Playing GOD is for general readers with an interest in health, self-help and spirituality. It delivers compelling recovery support for those 3–5% of us experiencing electromagnetic hypersensitivity (radiation pollution sickness). Or you may be one of the estimated 35% with mild to moderate symptoms, often without realising. I guide you through a practical radiation detoxification and recovery (DR RAD) program that can be incorporated into your – and your family and children's – everyday life.

Playing GOD explores the link between microwave radiation and spiritual effects. Do tower and device emissions disconnect us from intuition and universal Intelligence? How is radiation disturbing our minds and keeping us in an agitated state? It is a book which may expand your perspective in areas beyond electromagnetic radiation.

What sort of world do we want to live in? We might have an interest in co-creation of a society where health and harmony are priorities. Do we want to be dominated by technology or would we prefer technology work for us? Are wearable devices a prelude to transhumanism? What can we do to co-create a low-radiation future? **Playing GOD** talks to these topics and more.

In **Playing GOD** my journey with radiation pollution sickness, from decline to thriving health, is placed alongside important explanations. It was written to be read from cover to cover but if you're guided by intuition to jump from Part I to Part IV (or other) then do that! See the glossary if you are uncertain of any terms or definitions. The index provides another access point to specific topics.

Part I sets up a parallel context with cigarette smoking, which I use throughout the book. It explains three key needs affected by electromagnetic radiation – health, spiritual connection and freedom. Part I also details the 'empty being' perspective – the idea that we are made of electromagnetic waveforms, which is useful for concepts explored later in the book. It introduces new terms such as 'radiation pollution sickness' and 'cumulative electromagnetic radiation' as well as my journey to Sydney and the first stages of decline. I compare the body–mind to a symphony orchestra with each instrument an organ or gland.

Part II establishes the current status quo especially in relation to the very recent proliferation of microwave towers. It touches on derisive encounters with mainstream medicine and the paradigm (also in alternative medicine) of 'I'd like to see you again next week'.

Part II takes you through to the final disturbing stages of my Sydney EMF Experiment and the decision to move. I discuss some practically useful details such as telco tower trajectory patterns. I note the pattern of not learning from the past – are we repeating this with microwave radiation with our slowness to act despite the scientific evidence? The final section examines the blocks to co-creating change so we can see them within ourselves and dissolve them.

Part III documents a road trip out of Sydney – but the electromagnetic radiation exposure continues. I become a radiation refugee. I make it to a rural retreat but find Telstra is covertly wirelessly enabling. The history of exposure limits is briefly examined and thermal effect versus biological effect based limits are distinguished. I outline a chaotic ride smoking out my neighbour and consider the legal aspects of a neighbour's Wi-Fi. Our diminution of freedom along with a related probable decline in property prices are discussed as is the power of many to rise up.

Part VI takes to the air. What is it like flying with radiation pollution sickness? The concept of electromagnetic humans and the body–mind as an antenna is examined. The miracle that is DNA as a fractal antenna also makes it susceptible to damage. Both nuclear and electromagnetic radiation are deemed genotoxic at extremely low levels. Additional to synthetic electromagnetic radiation sources earth energies may adversely affect our health. I look at studies from Europe. The interactions EMF-EMF and EMF-Drug are introduced, and I talk about the 'master glands' and how our spiritual connection may be disturbed by radiation. The work of Rife with cancer and Lakhovsky with oscillation rates and his warning about the dangers of microwaves are briefly touched upon.

Part V begins on a snow and ice-laden trail in Patagonia, Argentina – mobile phones have their utility! I focus on the international and examine what is happening on Everest, in Mongolia and in Europe. Is there anywhere left for a radiation refugee to go? I also examine electromagnetic radiation and children and social norms around technology.

Part VI is all about mind control and microwaves as a weapon. I look at brainwaves and suggestibility and the way we are sold products. Is it time to go to plain packaging with mobile devices? A graphic example is provided. I talk about clicking 'voices in our head' and the danger of modulated frequencies (the physics of modulation is explained in Appendix D). The way microwave radiation can be compared with or even called a 'drug' is the subject of Part VI. I discuss 'weapons of mass destruction' – propaganda, mind control and dissemination of information to us as 'the masses'.

In Part VII I show you ways to recover from radiation pollution sickness. I guide you through DR RAD first aid transformation – Detoxify, Recuperate, Re-integrate, take Action and live your Destiny. You might decide to switch off with the I Quit My Mobile steps provided. There is guidance for preliminary internal work to collapse stress and trauma. I highlight how your radiation pollution sickness journey holds hidden gifts.

For two years electromagnetic radiation has been my life. I searched and experimented and correlated. My desire is that anyone on a similar or tangential journey of discovery spends their time on the planet in thriving health. We are electromagnetic humans and our body–mind is disturbed by electromagnetic sources. May this book inspire and empower you to co-create healthy change in your own life and the lives of your family and community.

You can reduce your personal electromagnetic radiation burden and enrich your family's health. Creation of a Zero EMF Sanctuary, a zone without the 'noise' of electromagnetic pollution, helps us reconnect to thriving physical, mental and spiritual health.

Together we can rewire society. Do we want to live in a country where our children are safe to go outside (and to school) without microwave-induced brain damage and behavioural changes? Are we going to watch the decline of the baby boomers to an epidemic of Alzheimer's disease and dementia that correlates with technology growth? Or will we see the evidence of multiple environmental triggers including electromagnetic radiation pollution? An achievable short-term target is reducing electromagnetic radiation exposure limits by a factor of 100 by 2017 and 1000 by 2020. Australia has the opportunity to step away from a paradigm of environmental illness and choose to value and prioritise health.

CHAPTER 1

A Pollution Epidemic

- *Mobile devices have replaced cigarettes as the new addiction, and they can be just as detrimental to human health*
- *Stealth technology – what telcos and mobile device manufacturers don't want us to know*
- *A new invisible pollution - electromagnetic radiation or electrosmog*
- *Current attitudes to electromagnetic irradiation of humans*

THE CIGARETTE–MOBILE DEVICE CONNECTION

A FOUR CORNERS PROGRAM ABOUT SMOKING AND the perpetuation of Big Tobacco reiterated that a highly addictive product used by a large number of people is a licence to print money. For decades the tobacco industry told bare-faced lies about the adverse health effects of their products. They lied knowing they were lying. The world's second largest cigarette company, British American Tobacco, recently announced a 24% increase in annual net profit to AUD$5.73 billion.[1] Its scientific director admitted that cigarette smoking kills. The program linked the company to shiny packaged products in Britain that target younger users.[2]

Mobile devices are a highly addictive product used by a large number of people. For decades mobile and telecommunications industries have told bare-faced lies about the adverse health effects of their products. They lied knowing they were lying. The World Health Organization has classified radiofrequency electromagnetic fields as possibly carcinogenic to humans (Group 2B).[3] Mobile manufacturers target market to children. Disney says 'use your smart phone to watch a Disney fairy…' and down the bottom of the webpage there are two girls that appear about three years old, dressed up as fairies and dancing.[4]

The Australian Communications and Media Authority (ACMA) 'is the government body responsible for the regulation of broadcasting, the internet, radio communications and telecommunications'.[5] The ACMA has administered the Australian 'Wild West' of telcos and mobile manufacturers with levity. They even have a marketing graphic 'citizenacma' seemingly encouraging children of around four years old to take up a mobile device.[6]

Why is there not more information on the adverse health effects of Wi-Fi, mobile devices and telco and NBN microwave towers? Why is the business so unregulated? It's because profitable telcos and mobile manufacturers don't want the general public to be empowered to rise up for the health of many over the wealth of a few. Business, government and so-called university experts on a payroll collude to enable an unfettered growth in public microwave exposure.

ELEPHANTS AND RATS – RADIATION IS EVERYWHERE

The elephant in the room is the topic that everyone knows about but nobody is willing to talk about. Manufacturers of man-made electromagnetic radiation are more like the rat in the corner cupboard. Not many of us know about them. The rat scoots around the skirting boards at 2 o'clock in the morning. Discreetly. It wants us to remain asleep. One icy winter in the wee hours we hear the rat rustling. Instead of getting up to chase it out, we roll over and press our pillow over our ears damping the squeaks. 'It's not that bothersome,' we tell ourselves. 'I'll buy a rat trap next week.' Meanwhile the quality of our sleeping and waking life deteriorates. As the rat gains confidence it makes more and more noise. Then one day we open our kitchen cupboard to see a rancid mess. That day came for me. A week after moving into an inner city residence insomnia and irritability arose. Headaches, brain fog, heart palpitations and depression soon followed. My interior cupboard was ravaged by telco and mobile manufacturer rats.

Mobile manufacturers and telecommunications companies are shrewd. They know our habits. Most of us don't know they are there. Telco rats use stealth technology such as hiding mobile phone antennas in the neighbourhood church cross. Antennas are camouflaged earth green and hidden behind pine trees in Byron Bay. Concealment doesn't reduce public exposure levels. Telco rats don't want us to know about Australia's military-derived radiation standards. They don't want us to know that other countries such as Russia have rejected our radiation exposure standards because they are dangerous to human health.

ELECTROSMOG, RADIATION POLLUTION SICKNESS AND THE SYDNEY EMF EXPERIMENT

We mostly hear about the dangers of chemical pollution through air, food and water. I lived in China for three years. Visibility was often reduced. It was like living 24/7 inside a man-made cloud. Often locals would walk the streets with a dust mask and occasionally I'd see someone with a full gas mask like those I'd attribute to chemical warfare. Beijing in particular had massive air quality issues. As a result a large portion of the population experienced respiratory disorders. Along with chemical pollution what other forms of pollution are present in our modern lives?

1. **Psychic pollution** – Psychic pollution can come from the TV news (societal), an angry relative (relational) or from our own internal generation of stress and anxiety (self).

2. **Electrosmog pollution** – Electromagnetic radiation induces physical and mental debilitation. Through actuating the cellular stress response it simulates the fight or flight reflex and shutdown. (Appendix A summarises common electrosmog sources.)

What if we are exposed to all of these pollutants as is inevitable in modern city and town living? What if we receive an acute dose of one of these pollutants? Cumulative and then an acute episode of electromagnetic radiation pollution exposure correlated with a deterioration in my physical and mental health beyond anything I'd previously experienced.

Playing GOD was inspired by a phase of life when I moved into an inner city apartment across from an antenna tower. Coinciding with increased radiation exposure my physical, mental, spiritual, vocational, social and financial health were wiped out. The human needs that were dramatically impacted during the acute phase of my radiation pollution sickness were –

1 **Health** – Current radiation exposure limits are based on military-derived thermal effects. In other words they are based on cooking flesh. Evidence indicates it is the information conveyed by electromagnetic radiation (as opposed to heat) that causes biological changes. These biological effects lead to loss of health, disease and death and occur at exposure levels hundreds of thousands of times lower than Australian limits.[7] I was exposed at levels well within Australian limits but correlated headaches, insomnia, irritation, anxiety and later heart palpitations, brain fog and depression.

2 **Spiritual connection** – I felt disconnected to my intuition and found it difficult to feel 'centred' in meditation. My experiences and those with clients indicate low level radiation exposure may contribute spiritual effects. Exposure may attribute to the spiritual disconnection and unease permeating modern society.

3 **Freedom** – After purchasing the apartment in Sydney and soon after getting sick, I had choices. I could move. I could spend tens of thousands to shield with uncertain results. I could stay on. Being driven from a city I'd just moved to was unacceptable. I stubbornly stayed on. When my health became utterly dire I had no choice but to move. I became a radiation refugee, excluded from towns and cities. When patrons smoked in pubs and restaurants we had a choice to socialise elsewhere if we didn't want to breathe cigarette smoke. With microwaves we have no choice. Electromagnetic radiation is everywhere from pubs and restaurants to botanical garden parks and beaches. Freedom is diminished.

At a dark point I decided to turn the losses and fraught journey into what I called the Sydney EMF Experiment. I sought to uncover every aspect of radiation and health in the Sydney EMF Experiment. My apartment and the city of Sydney became my science lab. I was simultaneously the microwaved chimpanzee and the observer. I documented where I was when I felt worse and what symptoms I was experiencing. I correlated the power of ambient microwaves in Bondi Junction versus the Botanical Gardens versus Bondi Beach. I noted my spiritual practice fade as symptomatic jitteriness made it difficult to sit still. I also spent thousands on products and advice as I sought to regain ebullient physical and mental health. With no magic cures I watched myself slump into a frightening despair.

The Sydney EMF Experiment simultaneously provided impetus. It was my driving force when so many other aspects of my life had deteriorated. I was determined to recover and share my journey and learnings with others to expand perspectives, inform and inspire action. The main themes coincide with requisite internal and external transformations to once again thrive individually and as a society.

STEALTH BOMBARDMENT I – NOW YOU SEE IT, NOW YOU DON'T

In the 1970s 45% of males and 30% of females in Australia smoked.[8] Current figures indicate around half as many are smoking.[9] Passive smoking kills 600,000 a year globally and nearly a third of those are children.[10] Government increased the tobacco excise by 12.5% in 2013 and will raise it by the same percentage over the following three years to raise $5.3 billion over four years. In Australia 'over 750,000 hospital bed days per year are attributable to tobacco-related disease and smoking has been estimated to cost over $31 billion a year'.[11] This approximately $124 billion in costs over four years makes the $5.3 billion over the same period look far less valiant. Extending the government treasury estimate over a decade suggests costs of $1.24 trillion to attempt to care for and repair Australians damaged by cigarettes. This is $1.24 trillion in financial aid to Big Tobacco.

The juggernaut rolls on. 'Altria Group plans to spin off its Philip Morris International tobacco unit, a move designed to give the overseas maker of Marlboros and other cigarette brands more freedom to pursue sales growth in emerging markets… The spin-off would clear the international tobacco business from the legal and regulatory constraints facing its domestic counterpart, Philip Morris USA.'[12] In other (my) words. 'It is getting harder to sell product. The serfs are waking up to cigarettes laden with deadly addictive chemicals. It took them a few decades to realise and even longer to enact change. We need to addict them earlier and focus on nations without interfering governments.' With love for humanity let's gently put a bomb under the CEO who earns US$20,139,967 a year by doing everything we can to have the share price go to $0.00.[13] We can check our investments. If we are capitalising from historically some of the most consistently soaring shares in the world then we are very much part of the pillaging cycle.

In 1997 Nick Xenophon ran for State Parliament on a no pokies platform. Australians lose $12 billion a year on pokies (2008–9) with a social cost to community of more than $4.7 billion a year. Up to 500,000 Australians are at risk of becoming problem gamblers and up to 5 million Australians are affected by problem gambling. Each loses on average $21,000 per year.[14] A watered down gambling reform legislation was passed in Federal Parliament. Xenophon described the reform as 'pathetic' and said, 'The poker machines industry is an industry based on exploitation and greed.'[15] Pokies aren't going anywhere. States are addicted to the $4 billion a year taxes they glean from gambling addicted Australians.[16]

Cigarettes and pokies cost the nation hundreds of billions of dollars while profiting industry. But we do have the freedom of choice to wreck our lungs with cigarettes and our finances with pokies. Electromagnetic radiation telecommunications now cost the nation undisclosed billions of dollars while profiting industry. We do not have a choice to have our DNA damaged by radiation. This damage is occurring at imperceptibly low radiation exposure levels.

Mobile manufacturer scientists have read the studies confirming biological alterations due to their products. They continue to be tight-lipped. As Big Tobacco did for decades, mobile manufacturers and telcos are taking the public for a quiet ride to the hospital and cashing in while they still can. Why do we allow them to continue doing this? One reason is radiation is invisible and we've been brought up with the dogma of Newtonian 'material' science. We are prejudiced to our sense of sight such that what we see 'exists' and can do us harm and that which we don't can do no harm. It is time to update our perspectives. We are empty beings and old-paradigm science of visible matter captures a miniscule fraction of reality.

CURRENT ATTITUDES TO STEALTH IRRADIATION

People fit into three groups in the matter of stealth irradiation of humans –

GROUP 1 – Want more towers. They may have a vested interest (telco shares, job) or are technology infatuated driven by faster streaming and download speeds no matter what.

GROUP 2 – Are ambivalent. They are suspicious of decision-making authorities and feel they don't have a voice on the topic. Other life issues take priority. Ultra high download speeds on their mobile devices are unnecessary. They may have experienced minor adverse effects of technology rather than serious debilitation.

GROUP 3 – Don't want more towers. They've either experienced their own health decline or that of a friend or family member correlating with technology exposure. They doubt industry and government motives and actions and are activated to create a healthier society.

GROUP 1 is the minority group, but radiation regulations support their technocratic paradigm. GROUP 3 is expanding and GROUP 2 is the ambivalent majority only needing empowerment through knowledge to jump across to GROUP 3 and activate.

A client, Alex, came to me for restlessness, sleeplessness, anxiety, irritation and headaches. He'd heard about my trauma collapse work and thought it might be useful. As a member of GROUP 2 he saw technology as inevitable and 'out of my hands'. I'd experienced Alex's symptoms myself. They are common symptoms attributed to electromagnetic radiation exposure and unresolved stress and trauma. Alex told me he used his phone for over an hour a day and had Wi-Fi in his home office. The modem was on his desk barely half a metre from his study chair. I asked him to journal his days and correlate his symptoms with time of technology use.

A week later we met and he told me about his epiphany. He was getting his worst symptoms in the evening in his home office and this was when he'd call his girlfriend. He'd go straight to bed minutes after being buzzed by the computer screen and his Wi-Fi remained switched on during the night just metres from his bed.

We set him up with an acoustic tube headphone for his calls. He hardwired his home office. He switched off all devices an hour before bedtime. Alex's symptoms declined. On a work trip he was assigned a hotel room directly across from an antenna and was empowered to ask for another room. Alex went from GROUP 2 to GROUP 3 in a period of two weeks. I had a similar call to action and shift to GROUP 3 during the Sydney EMF Experiment.

CHAPTER 2

The Fabric of the Universe

- *Universal ether and the life energy that connects all things on Earth*
- *Susceptibility to radiation pollution sickness and the four E's – empathy, exposure, emotions and the environment*

EMPTY BEINGS AND THE HOLOGRAPHIC CONCEPT

IF WE SQUEEZED ALL THE SPACE OUT of matter the entire human race, all seven billion of us would fit into a sugar cube.[17] We are 99.999% empty. NASA states that over 95% of energy density in the universe has never been detected in the laboratory.[18] The science of matter we have been taught explains only a tiny fraction of reality.

Quanta are the smallest physical part in an interaction at this time. Physicists have determined that the only time electrons and other quanta manifest as light is when we look at them. Otherwise they are waves. We look at our partner and see a good-looking physical reality with a three-dimensional volume and a variety of textures, colours and hues. If we look away our partner is a multitude of interfering waveforms. Our viewed reality is based on the particle or light nature but behind the scenes it is a sea of waveforms of varying frequencies. These waveforms criss-cross and interfere in an infinite mesh and are filtered through a holographic unit (our body–mind) to create what we think are objective perspectives. Life is like watching a projected movie.

I'd been out of touch with a good friend for many months. One day I felt a building sense to contact her. Finally the impulse to connect was so strong that I sent her an email to ask if she was okay. I had no idea that she was in Guatemala and it was early morning there. She replied saying, 'Wow! Thanks, Ben. I've got nasty food poisoning. You emailed me at the exact moment I started throwing up.' I don't have a sense to email or call every time a friend has an upset stomach or stubs their toe. There was an intelligent force involved also. My friend had been experiencing the travel doldrums. She said that between vomiting episodes she cried when she received my email and it gave her strength to carry on with her travel plans. After returning a few months later she said was pleased she didn't give up on what turned into a transformative adventure. Cosmic intelligence and what David Bohn called the non-locality connection or 'spooky action at a distance' were involved to nudge me into contacting her at that moment. Similarly we have an impulse to call someone and just as we press the digits into our phone they call us.

David Bohm's quantum non-locality analogy involves a fish swimming in a tank. We have never seen the fish or tank before. The only information we have is from a video camera at the side of the tank and another at the front. The front perspective comes up on a screen in front of us and the side perspective on another screen. The angles are different so viewing the two screens we might say there are two fish being projected. Then we notice when the fish on screen #1 moves there is a corresponding movement of the fish on screen #2. They are either instantaneously communicating with one another or they are the same fish. If the fish are one and the same then no communication is taking place. Non-locally connected particles are part of the same organism.[19] When one part moves, through non-local connection another responds instantaneously.

The holographic and intersecting waveform mesh perspective will be useful to return to as we explore the non-thermal effects of electromagnetic radiation in later chapters. What is the medium for this instantaneous 'spooky action at a distance'? What fluid carried the energy of my friend's distressed situation in Guatemala to me on the east coast of Australia and an urgent impulse to contact her?

MESMER AND THE FLUID THAT CONNECTS

Once upon a time I had the family dog Chilli stay with me. I started the day with a list of ten things to do. Two days a week I worked with clients on trauma, guilt and shame release and upgrading their beliefs. My first client cancelled due to family illness. I went off to update car registration but didn't have sufficient identification. Later in the day I arrived at the bank as they were locking the door. Then the hardware store had sold out of the building products I required for some renovation work that evening.

Heading home for solace I pulled into the driveway and Chilli raced out. Like a sprint car on a racetrack he stretched the laws of physics with six loops of the front yard, tearing up the grass in the process. He then splayed belly-up at my feet. I laughed and scratched his chubby torso. He jumped up and did another six loops. He was often pleased to see me but this excitable racing around the yard was unprecedented. My sense was that Chilli dog knew I'd had a rough day and he was cheering me up! How did Chilli know?

Franz Mesmer called the life energy that connects all things a 'fluid'. He proposed –

- A responsive influence exists between the heavenly bodies, the earth, and all animated bodies.
- A fluid universally diffused, so continuous as to admit no vacuum, incomparably subtle, and naturally susceptible of receiving, spreading, and communicating all motor disturbance, is the means of this influence.[20]

Wilhelm Reich called this tangibly measured life energy 'orgone' (from life energy of orgasm). It relates to Chinese feng shui concepts that are thousands of years old and India's equivalent ancient science Vastu Shastra.[21] The basis for Vastu are the five elements of earth, water, air, fire and space. According to Vastu the human body and the physical world are made up of these elements and life is sustained because of their balance. The five elements relate to the five senses. Space or the universal ether is linked to sound and our sense of hearing.

The universal ether is also associated with thought transmission. As thought waves move through space we pick up on harmonious or disharmonious thoughts of another person. The universal ether is the medium for harmonious electromagnetic radiation transmissions produced by nature and disharmonious synthetic transmissions from technology. Why is it that some people seem to be affected by these transmissions more than others?

ELECTROMAGNETIC HYPERSENSITIVITY – THE FOUR E'S

In my first years of life there were numerous social outings with the adults. The sailing club was a regular gathering. The old man was a champion yachtsman. He'd often win and raucous festivities ensued. When I wasn't playing with other kids I'd hang out in the crowded bar. I stood readied for action like a ball boy. Rather than retrieve tennis balls I'd order jugs of beer to take back to the crew. At five years old I had to climb on a bar stool to order. Occasionally a grown-up would hoist me onto the stool. At times I amused the adults by taking a sip of beer but I never allowed frivolity to hinder the return of a full jug into the drinking circle before glasses were emptied.

Beneath the celebrations dark things were taking place. I was too young to fully understand but I knew that men and women were going with women and men that were not their wives or husbands. Sometimes there were tears. Occasionally there was yelling. More often it was subtle energy (or thought transmissions) that I'd pick up. Through Mesmer's fluid that connects all things I picked up disharmonious transmissions of fear, jealousy, sadness and rage. This medium also informed me of harmonious love and joy in which case it felt safe to interact.

I occasionally incurred the wrath of the adults. Due to my openness I'd sense things the adults thought nobody else knew about. I asked Mrs Dougherty, 'Why are you scared of Mr Dougherty?' or 'Why don't you want your daughter anymore?' Initially I had no social understanding and naively asked these questions in front of everyone. Exposed truths stirred their emotional pot and fury would ensue. After a couple of rounds of wrath a survival decision was made to stick to my role of making sure nobody had an empty glass. It became dangerous to utter my observations so I ceased speaking but I continued empathising with the pain of others.

Disposition to radiation pollution sickness (or electromagnetic hypersensitivity) is more likely if we are one or more of the four E's –

1 **Empathic** – I've observed in myself and others that a sensitivity to other people's emotions (including psychic pollution) parallels sensitivity to electromagnetic radiation pollution. I was empathic at the sailing club at six years old and even though I tried to shut this down I continued to be empathic 20 years later. Since then I've learnt to remain sensitive without continually 'taking on' the energy

of others. Part of this journey was reviving an underactive third (solar plexus) chakra as detailed in Part IV – Spiritual Effects, and Part VII – Recovery.

2 **Exposed** – Electromagnetic radiation exposure leads to cumulative electromagnetic radiation (CEMR) damage. For many years I worked with technology. At one job I had three computers and eight screens. I flew internationally a few times a year and domestically every other week. All of this led to increased vulnerability to tip into decline. Twenty years ago I did not have a significant cumulative electromagnetic radiation burden and therefore it is unlikely I would have deteriorated so dramatically after exposure. Cumulative electromagnetic radiation exposure is a reason pilots and flight attendants are highly susceptible to radiation pollution sickness.

3 **Emotional** – Held emotions become stress and trauma psychosomatically 'locked-in' to our system. Cells are then required to respond to our internal stresses as well as external toxins including radiation and chemicals. In fight–flight they are not able to detoxify and 'breathe' and eventually they shut down.

FIGURE 1 – THE ELECTROMAGNETIC SPECTRUM[22]

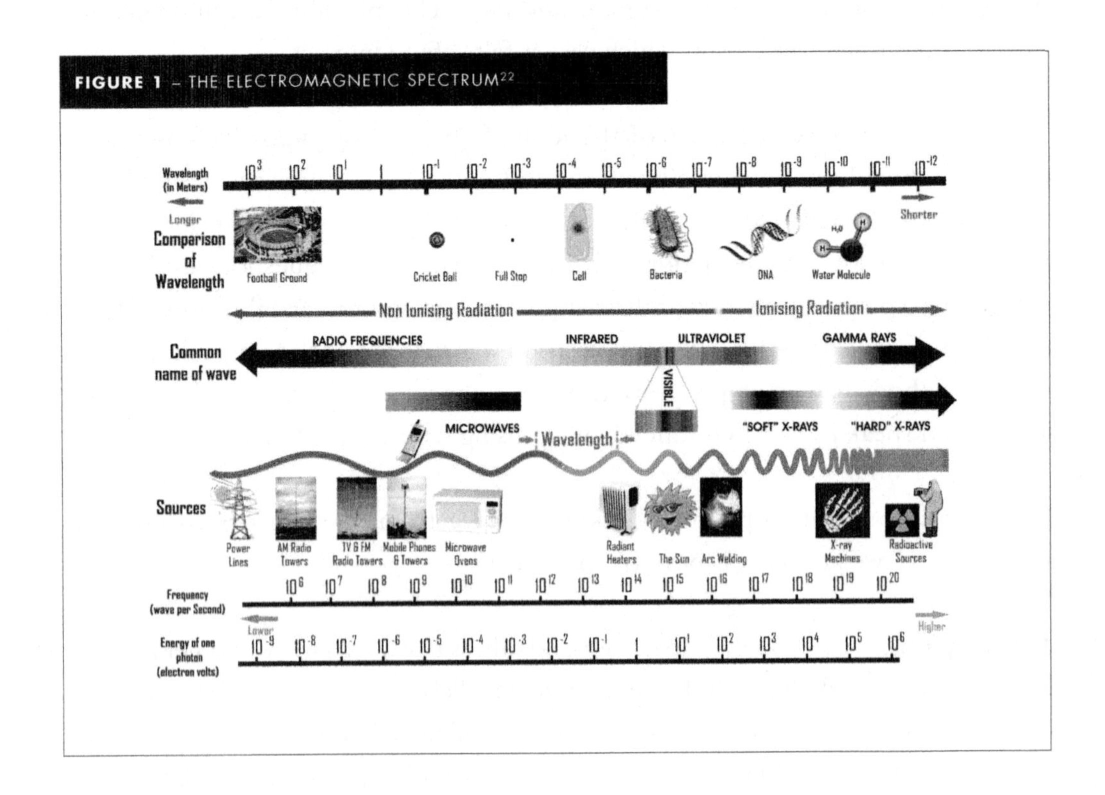

4 **Environmental** – Exposure to toxins such as mould exposure or chemicals is another stressor. Cells are overwhelmed with multiple stress fronts to deal with. Chemicals, yeast infections, heavy metals and parasites may contribute. The decline of radiation pollution sickness has parallels with multiple chemical sensitivity.

I prefer the term radiation pollution sickness (RPS) because it captures the reality of electromagnetic hypersensitivity as an environmental illness. The word 'radiation' alarms the public to the truth so industry and government agencies seek to minimise the use of it in descriptions. They substitute the dishonest but less threatening 'energy' or 'radio-frequency' to describe radiation. Non-ionising radiation from microwave towers and mobile phones is radiation from the electromagnetic spectrum (Figure 1) as unambiguously as ionising X-rays and the nuclear disasters Chernobyl and Fukushima. (Appendix D explains ionising versus non-ionising radiation physics.) Evidence shows exposure to either form of radiation causes DNA damage.

CHAPTER 3

Introducing Cumulative Electromagnetic Radiation

- *How electromagnetic radiation builds up in the body*
- *Radiation experiments on chimpanzees*
- *The continual onslaught of electromagnetic waves from telco towers emitted at multiple frequencies*

THE RASPBERRY CORDIAL ANALOGY

IF RASPBERRY CORDIAL IS ELECTROMAGNETIC RADIATION LET us say pure water in a jug is our clean energy field. During recuperative sleep, through nutrient conversion and prana, pure water is constantly being added to the jug. A few drops of raspberry cordial pollute the water but our body–mind recuperates by adding pure water until the jug is once again clear. A single acute exposure or chronic low intensity electromagnetic radiation exposure may result in a jug with the number of raspberry cordial drops outnumbering the number of pure drops. Our body–mind does not keep up and soon the jug turns red. At a point the burden of cumulative electromagnetic radiation exposure becomes so great that our body–mind can no longer add sufficient pure water to detoxify. It shuts down. What were once minor symptom-free exposure events now induce mild, moderate or severe radiation pollution sickness symptoms. Suddenly all it takes to induce a headache is being in a room with people with mobile phones switched on.

Cumulative electromagnetic radiation is our non-ionising radiation exposure from our conception to this point in time. It includes radiation from every mobile phone call we've made and every night we left our Wi-Fi running as we slept. If we carried meters around with us 24/7 we would see a spike each morning when we arrive at our wireless enabled office. We'd notice a jump in levels at the café as we chat with a friend on our phone and another peak level as we sit in an aeroplane waiting for take-off. Even in our homes microwave radiation levels may show millions of times greater than those of 20 years ago.

Military-industrial complex proponents of a thermal effects basis for non-ionising radiation exposure limits understood the danger of repetitive exposures when they said – 'It should be understood that a cumulative effect is the accumulation of damage resulting from repeated exposures each of which is individually capable of producing some small degree of damage. In other words, a single exposure can result in covert thermal injury, but the incurred damage repairs itself within a sufficient time period, for example hours or days, and therefore is reversible and does not advance to a noticeable permanent or semi-permanent state. If a second exposure or several repetitive exposures take place at time intervals shorter than that needed for repair, damage can advance to a noticeable stage.'[23]

STUDY OF CUMULATIVE EFFECTS ON CHIMPANZEES

A biophysicist at Northrop Space Laboratories in California prepared a paper called 'Biological Entrainment of the Human Brain by Low-Frequency Radiation'. If the body's biological clock was related to electrical impulses called alpha rhythms then could electromagnetic fields alter that clock?[24] The file went to the acting director of the Advanced Sensors program of the Advanced Research Projects Agency. Project Pandora was initiated to discover whether very low level microwave signals could direct the mind.[25]

A research facility was assembled and a signal of 2.2–4.0 GHz with power levels up to 5mW/cm^2 generated. These were exposure levels lower than the US limit 10mW/cm^2. (Appendix D explains electromagnetic radiation measurement units.) The chimps had been trained to perform specific tasks. After the first week of 10

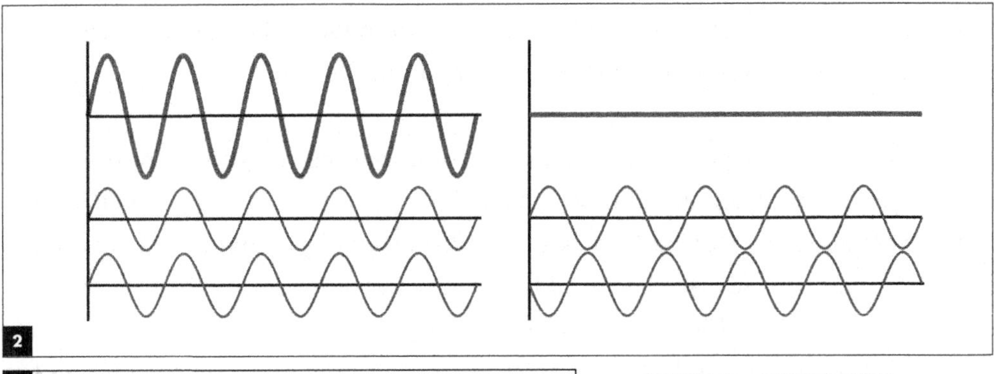

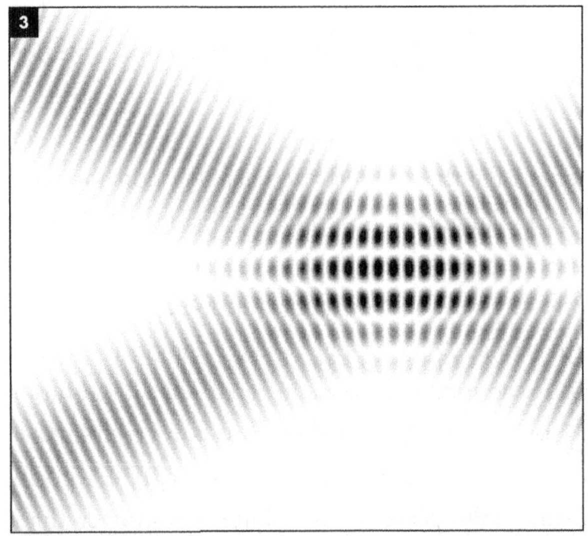

FIGURE 2 – CONSTRUCTIVE INTERFERENCE (LEFT) AND DESTRUCTIVE INTERFERENCE (RIGHT)[28]

FIGURE 3 – CONSTRUCTIVE WAVEFORM INTERFERENCE MAP[29]

FIGURE 4 – WAVEFORM ATTENUATION[30]

FIGURE 5 – WAVEFORM ATTENUATION AND INTERFERENCE[31]

hours per day exposure the chimps successfully completed the tasks.

On the twelfth day a slowdown of the chimp's work function occurred. On the thirteenth day the test animal stopped working and appeared to be in a deep sleep for two days until the radiation was turned off. It then returned to the work function over the next five days. The radiation signal was again switched on and this time the chimp slowed down in eight days and stopped and went into a deep sleep. They left the chimp like this for three days and when they ceased irradiating it did 'not return to normal'.[26]

Given the results the new code name became Project Bizarre. Researchers noted 'potential of exerting a degree of control on human behaviour by low-level microwave radiation' and that the US needed to overhaul radiation exposure standards to take into account 'non-thermal damage potential' or biological effects.[27]

The chimp was no Academy Award winning actor. During the Sydney EMF Experiment I observed the same cumulative weakening of systems. When away from radiation I recuperated. After returning to the radiation exposed apartment and inner city my systems deteriorated more rapidly than they had previously. By the end of the Sydney EMF Experiment symptoms that had once (when I first moved to the city) taken a week to surface instead took hold within hours.

RADIATION SOUP

When we arrive early to a party the music is low and only a few people have arrived. We converse with a friend at 'normal' volume. Later in the evening the crowds have arrived and the music has been turned up. Sound waves generated by the voices of party-goers and the music interfere and we must speak loudly to be clearly heard.

Similarly electromagnetic radiation waveforms interfere. Figure 2 shows two waveforms constructively interfering to generate a 'stronger' waveform at the top. When two waveforms are out of phase by 180 degrees as in the second example in Figure 2 the waveforms destructively interfere to cancel each other out as represented by a flat line. Figure 3 uses intensity mapping to show an increase in amplitude or intensity (darker = higher amplitude) where the waveforms intersect.

Another approach is to imagine the drop in the lake in Figure 4 as an electromagnetic radiation point source such as a mobile phone tower, Wi-Fi modem or mobile device. The lake is the universal ether surrounding us. The induced signal is strong close to the power source (high amplitude wave fronts) and attenuates further from the source. Telcos remedy this attenuation by installing networked towers to create full coverage.

In Figure 5 multiple drops have landed in the lake. Each drop is the equivalent of a radiation source. The wave fronts attenuate as they spread further from the point source and interfere with wave fronts from other point sources to create waveforms of variable amplitude and frequency. Each telco has its own frequency bands and associated antennas to emit a specific frequency of radiation. For instance a telco tower may contain antennas owned by Optus and Vodaphone emitting at multiple frequencies. Imagine now the lake has hundreds of drops creating their own waveforms. The lake becomes a chaotic electromagnetic radiation soup. Some waves destructively interfere and cancel. Others constructively interfere and amplify.

We are a point somewhere in the lake attempting to process this radiation noise. When our body–mind is in a suboptimal state or if the radiation onslaught is continuous it may break down. When we swim in a pristine lake we leave feeling refreshed and vibrant. When we swim in a lake contaminated by multiple chemical pollutants our health might be adversely affected. As with radiation pollution sickness the health effects of our swim might take a while to become evident. If we have just swum in a polluted lake and are from a vulnerable demographic such as those aged or hospitalised or developing infants and children we may be more susceptible to decline. If we are in the lake continuously 24/7 no matter how optimally our body–mind was functioning or what demographic we are from we will be adversely affected by pollutants.

CHAPTER 4

My Journey I – Sleepless in Sydney

- *A seemingly idyllic move to city living*
- *Mysterious symptoms and ill health begin to take over*
- *The six criteria used to diagnose radiation pollution sickness*
- *The reality of my illness, and telling my family and friends*

COURTING SYDNEY

THE PULL TO MOVE TO SYDNEY WAS overwhelming. Suddenly I was daydreaming about creating a life there. I did calculations, searched online real estate and talked to people about work opportunities. The specific area of interest was Potts Point, which was a suburb I'd never previously set foot in.

I took reconnaissance weekend trips to the area. The weekend farmer's market was a drawcard and I loved the walkability and community buzz. The suburb was also booming and an investment might be a financial win. At an open home I was walking through a studio space with 30 potential buyers. I spotted a sharply dressed woman who captivated me far more than the apartment did. Walking out of the building I saw her leaning on the stairwell railing talking to another lady. I playfully asked her to check my 'broken' ribs after getting a double elbow from an elderly couple during the inspection. She'd been a bit bruised in the mosh pit also so we brainstormed a team-up strategy to eliminate the competition. Ideas included placing malodorous objects in the bathroom cupboards and gluing 'Inspection Cancelled' stickers over the 'Home Showing' signs at the building. Katie gave me her phone number and we went off to our selected apartment inspections.

I flew home and texted her a week later. We learnt about each other via text messages. It was fun and romantic courting this way. Mobile phones are the bee's knees for fun and romance. We spoke on the phone. We saw each other a couple of times and it was even better in person. There were city-girl things I couldn't quite fathom like visiting the gym every other day and having a non-stop busy life. It was clear I was falling for her. My mobile phone assisted in that falling. It also assisted me practically. I travelled to Sydney searching for an apartment and my mobile phone navigation and internet enabled eight apartment inspections in one day. Via my phone I could reschedule a flight to spend longer in Sydney with Katie.

When we were away from each other our mobile phones were a lifeline. We had a way to connect. Sometimes it felt even better than being together in person. There was none of the messiness. I'd text her at home as I ate dinner on the deck. We'd remotely share a romantic meal together. 'I've just opened a chardonnay. Wish you were here to taste it with me.' We photographed our favourite meals and indulged in a small-scale selfie-fest. Texting allowed seconds and minutes to think of a response that I may not have come up with in person. Virtual Benjamin was wittier than the real one. Katie was a smart girl in person but it was that initial texting phase where our creative energies peaked. Texting was a platform for poeticism and Katie was so proud of our text chain she suggested we make it a book of poems.

I found the 'ideal' apartment. I had tabs on an 'ideal' job. I was head over heels with my 'ideal' mate. Through texts, selfies, emails and occasional voice calls the fantasy amplified daily. My external world was a glossy picture book of success and I was running so fast I didn't notice the quiet badgering of being off Path.

THE DECLINE

The big smoke beckoned and I answered the call. Katie met me at the airport. She'd obtained the keys and wrapped curly purple ribbons around them along with purple gifts, cushions and candles. We smudged the apartment with white sage and sounds of the didgeridoo. Inner city hippy stuff. The neighbours knew there was a new kid on the block. The first days were glorious. I met Katie's relatives and liked them. I met a green roof designer and started planning what would have been one of Australia's first inner city rooftop food forests. The days were inspiring, flowing and full of new social connections, innovation, pizza joints and ambling hand in hand late night gelato excursions to balance Katie's habitual exercise.

I squeezed a lot of life into those handful of days but within a week I felt off kilter. Everything in life was working out externally yet I felt agitated. Considering I was 'living the dream' I couldn't fathom why the unrest. A constant headache soon became a 24/7 companion. This was a new experience for me. Insomnia kept me up most of the night leading to groggy, unproductive days. Occasionally I'd feel a tingle on my tongue and pins and needles at my extremities. At first I assumed it was due to the change in environment but I'd lived in major cities and travelled and moved home dozens of times previously. Why now? The fierceness of the symptoms concerned me.

A few weeks after moving into the apartment heart palpitations began. Mental health disturbances amplified from agitation and jitteriness to brain fog and depression. There was a five stage pattern of decline –

1 **Denial** – Initially symptoms including anxiety, headaches, insomnia and fatigue were relatively lighter. I had no idea at first that electromagnetic radiation may be a factor. Radiation was not something I ever thought about. I attributed decline to other sources saying, 'I must be stressed' and 'This is temporary as I am adjusting to a new city.' Life was not particularly stressful though. I knew there was more to it but electromagnetic radiation was not on my radar as something that could topple my health.

2 **Frustration** – I searched for but could not find why I'd suddenly taken ill. I considered every possible contributing factor before considering electromagnetic radiation. In this stage people will typically seek medical advice and be left frustrated and without a diagnosis. I waited until Stage 4 to see a doctor. People may remain in this stage for years due to limited knowledge and awareness around adverse biological effects of electromagnetic radiation. I was fortunate to decline so rapidly and noticeably when around technology

that I was rapidly led from Stage 2 to Stage 3 and a clear correlation.

3 **Correlation** – During my decline I had a string of days outside of Sydney where I noticed I slowly improved. After three days in a rural setting I felt healthier and my sleep improved. When I returned to Sydney I rapidly deteriorated. This is when I stepped onto my rooftop deck and tilted my head up to observe the adjacent antennas. The link between my symptoms and electromagnetic radiation emitting technologies became undeniable. I'd discovered a replicable cause which was confirmed over and over. My curiosity turned into a two-year journey of research and writing and a four-month immersion of self-experimentation I named the Sydney EMF Experiment.

4 **Loss** – I was desperate for a solution and sought traditional and alternative remedies. 'I need a magic pill to make this go away.' Nothing delivered. 'I want my old health and life back.' With extreme anxiety came susceptibility to marketing promises and I spent thousands of dollars seeking solutions. Life events in Stages 1–3 compounded and I lost my job, relationship and later my home, assets and investments. Add to that physical and mental health and some days it felt like I'd lost it all. The experiment kept me going.

5 **Depression** – When I'd tried everything and did not find a 'hopium' pill for radiation pollution sickness my thoughts became despairing. 'If I am to live like this with 24/7 symptoms I'd rather not live.' I was cornered into leaving the home I'd just purchased.

I went from Stages 1 to 5 in four months. My health dropped to a two out of ten. Once I moved to a Zero EMF Sanctuary for recovery Stages 4 and 5 persisted with further financial losses and depression. My physical health returned to a five out of ten but the four months of acute exposure had left residual mental health scars. I was also at the tipping point where symptoms were easily triggered. When I was in a crowded room for ten minutes I immediately got headaches from passive mobile radiation exposure. Catching a short flight would catapult my health back to a two out of ten. Agitation, headaches, palpitations and brain fog returned rapidly with exposure such that flying or being in a city became impossible. The last time I was at an airport I was so repulsed by being in that environment that at the gate I turned around and left as my flight boarded. My freedom was diminished and I couldn't get on the plane, missing another important event.

DIAGNOSING RADIATION POLLUTION SICKNESS

During my phase of decline I visited three doctors, two psychologists and had my first visit to a psychiatrist. At one appointment the GP said flippantly, 'You're being overly sensitive. You need thicker skin.' Our ten minutes were up. I left empty handed and slightly derided. If I'd had a little less brain fog I might have responded, 'Thicker skin does not reduce the penetration of electromagnetic radiation into my skull, thyroid and parathyroid glands.'

'Hypersensitive' according to the Oxford Dictionary is 'abnormally or excessively sensitive'.[32] Electromagnetic hypersensitivity is therefore a condition label suggesting patient dysfunction. It is a disempowering label which diminishes the patient. The doctor who said I was 'overly sensitive' might have instead observed, 'Your system is being battered by environmental pollutants.' It is not patient dysfunction but rather a radiation polluted environment causing the debilitation. In Sweden they call electromagnetic hypersensitivity an impairment. Olle Johansson from the Swedish medical university Karolinska Institute has said, 'No human being is in itself impaired, there are instead shortcomings in the environment that cause the impairment. Thus, it is the environment that should be treated!'[33]

Labelling electromagnetic hypersensitivity (EHS) as a 'condition' also forms a superhighway for Big Pharma drug creation. I was prescribed an antidepressant for my symptoms. Prescribing neuroactive medication for electromagnetic radiation or chemical sensitivities is like putting a band aid on a broken limb. The core issues are not being addressed. Watch out for a patented EHS drug coming soon to a pharmacy near you!

Like humans, great white sharks are electromagnetic creatures. Within a few hours of being placed in an aquarium a great white often dies. Researchers believe electrosensitivity is the cause of death. A 2.3 metre juvenile great white, 'Sandy', was kept in captivity for around three days. She kept banging her head on a certain part of the tank. Aquarium director John McCosker and colleagues searched for and found a miniscule potential difference of 0.000125 volts in that area.[34] Sandy's senses were disturbed by an electrical field and her health was adversely affected.

Environmental medicine expert Roy Fox believes there is an underlying change in the nervous system in both electromagnetic hypersensitivity and multiple chemical sensitivity. He suggests using the same six criteria identified for the diagnosis of multiple chemical sensitivity for radiation pollution sickness.[35] Episodes may be triggered by an acute or chronic exposure. Comments on my own case follow each criteria –

1 **Symptoms are reproducible** (with repeated exposures) – When I left Sydney and went to the countryside my symptoms declined. When I returned to the city symptoms returned. At a large event I was ill. After leaving the event symptoms declined.

2 **The condition is chronic** (has persisted for a significant period of time) – I had four months of severe decline followed by detoxification. After this period there were occasional severe episodes when I put myself in a high exposure environment.

3 **Low levels of exposure** (lower than previously or commonly tolerated) result in symptoms – Initially I was unable to stay in hotels with their Wi-Fi. I was unable to attend events due to passive mobile device use. I was unable to be in the city or an office with Wi-Fi. The radiation near antennas was often intolerably strong. Symptoms would even arise at beaches with nearby (< 1 km) towers. The majority of activities we take for granted I was unable to take part in.

4 **Symptoms improve or resolve completely when triggers are removed** – Symptoms improved significantly when I moved to a Zero EMF Sanctuary. I left Sydney and moved to the countryside – a remote area so far from towers that mobile reception was intermittent to negligible. Even in this environment I experienced ongoing residual symptoms and trips to town were initially challenging. (In later chapters I discuss why it's important to not only remove the external triggers but also to release internal stresses for full resolution.)

5 **Responses often occur to multiple unrelated triggers** – My symptoms worsened on one occasion when I was exposed to mould in parallel with electromagnetic radiation triggers. On another occasion a stressful life situation amplified symptoms disproportionately.

6 **Symptoms involve multiple organs** – I experienced brain fog, heart palpitations and tingling in my tongue and extremities. Glands including the pineal and pituitary were affected as I discuss in later chapters.

SENSITIVITY, EMERGENCE AND SHARED EXPERIENCE

As I found myself developing an aversion to airports, shopping malls and Wi-Fi enabled cafés I also noticed enhanced sensitivity to noise, emotions, scents, touch, mould and chemicals. Electromagnetic radiation exposure was not the primary driver but rather a small piece in this increased connection to the senses.

Mostly I kept to myself. When I tentatively discussed my ill health with family I didn't get the blank stares of the medical profession. Family listened intently; however, they were mobile device users too – and healthy. How could something invisible that everyone uses catalyse a deterioration in physical and mental health? Where did the energetic and robust Benjamin go?

At first if the topic came up I was hesitant to discuss it. People were nervous about what microwaves were doing to their health but they 'didn't want to go there' to explore the topic. They pushed the possibility of radiation pollution sickness out of their minds and had enough life worries with bills, family, work and holidays to arrange.

Invisible electromagnetic radiation is shrouded in science and we have been programmed to doubt our ability to understand it. After being a radiation refugee for many months I came out of hermitage and met with friends. I found myself conveying the essentials of electromagnetic radiation with simplicity, clarity and positivity. The disempowering scientific mystique created by so-called experts needed to be dismantled.

The Sydney EMF Experiment was an immersion which provided insights. My friends wanted to learn more. A couple of them had children and wished their kids the healthiest life possible. We implemented immmediate actions. Another sought guidance for the internal transformation that I talk about in later chapters. So it started. I was a resource with a unique perspective to share. One of the analogies I used to convey truths during that early stage was that of our body–mind as a musical masterpiece that can also be out of tune and noise disrupted.

CHAPTER 5

Waveforms and Vibrations

- *How our body responds to the noise of microwave radiation similarly to the way we respond to discordant music*
- *The Hum of the universe – and how radiation pollution alters the vibrational signatures of where we live*

BODY–MIND AS A SYMPHONY

ONE NIGHT I WENT TO THE LOCAL hall to see an acoustic band. The relationship between notes was exquisite and resulted in a delightful sound being communicated. What if each of the band members had been out of time with each other by a second or two? What if each member had played a completely different song with the guitarist playing track 1, the pianist track 5 and the singer on track 10? What if each musician tuned to a different 'key' or frequency? It would sound noisy rather than musical and the audience would soon leave. We innately understand rhythm and harmonious relationships as indicated by common language. 'She was on my wavelength.' 'That really resonates.' 'We were in-sync.'

Figure 6 shows two waveforms out of phase by Θ. This could be someone saying 'ha' then again 'ha' a moment (Θ) later so we hear 'haha'. When phase shifted Θ to overlay the waveforms are in-sync and resonate as one to coherent 'ha'.

Each part of our body–mind oscillates at different frequencies. Cellular frequencies differ from brainwave frequencies and the frequency (or rate) of our heartbeat. Yet, as in a magnificent symphony orchestra, coherent relationships exist. In an orchestra the bass drum, tenor drum, violin, harp, piano and clarinet all have different sound frequencies but play together harmoniously. Healthy DNA strands, cells and organs are electromagnetically engaged and play together with perfect harmony.

Sickness or fearful emotional states internally generate discordance. We are at a family Christmas dinner when a drunk uncle blurts out something offensive and the harmony of the gathering is shattered. What if the harpist in our symphony analogy is a diseased liver and unable to play in tune? Other instrumentalists in our body-mind symphony are affected and the overall (holistic) 'sound' of our body-mind is noisy rather than sublime.

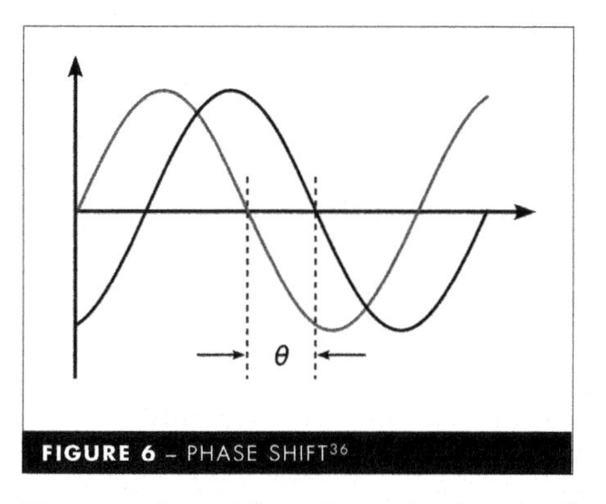

FIGURE 6 – PHASE SHIFT[36]

Our body–mind generates non-linear sinusoidal waveforms per Figure 6. Nature is based on non-linearity. Yet via technology we are now exposed to linear (rectangular) pulsed modulated electromagnetic radiation waveforms (such as Figure 13 in Appendix D). Pulsed modulation has been shown to induce agitation and mind alteration as outlined in Part VI – Weapons of Mass Destruction. Therefore the level of synthetic microwave radiation exposure is of public concern as is the type (or form) of transmission. Linear pulsed telco signals at relatively low power density may cause significant adverse biological effects.

Consider each of the six waveforms in Figure 7 to represent an electromagnetic radiation source at a moment in time. One waveform might represent an Optus tower signal, another a Telstra antenna emission, another a TPG antenna emission and the final three Wi-Fi hotspots and mobile phones. It is noisy. How noisy would it be if there were one hundred or one thousand waveforms per everyday life? A fraction of a second later the amplitude, frequency and phase of the waveforms shifts. We start talking on our mobile phone or we walk into direct line of sight with an antenna and the waveforms change. Only something as perfectly designed as our body–mind has a chance of coping with the fluctuating noise. The conductor of a symphony orchestra would throw her hands up in the air after seconds of the noise our body–mind endures.

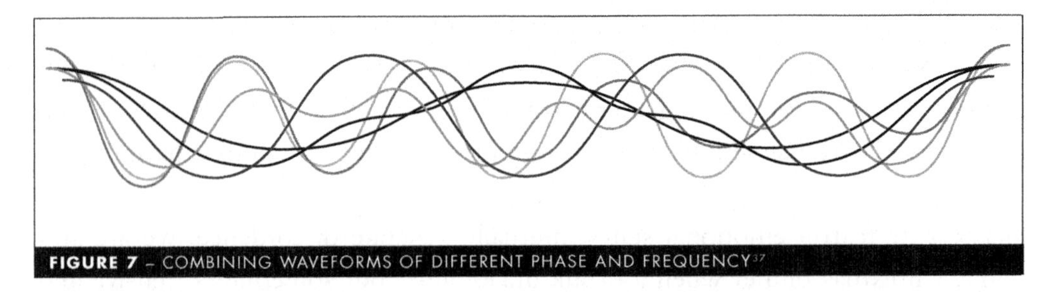

FIGURE 7 – COMBINING WAVEFORMS OF DIFFERENT PHASE AND FREQUENCY[37]

At Dharam House our focus is on empowering by passing on knowledge related to electromagnetic radiation and health. We do this through tailored talks to groups including parent and teacher school groups, local councils, community groups and businesses. We are honours graduates, experienced as health practitioners with Certificate IV Training and Assessment qualifications. We share our unique insights, research and personal experience with radiation pollution sickness combined with our clients' experience. During talks we sometimes demonstrate the sound of multiple source microwave radiation noise using a meter. In a remote national park the meter is quiet. Inside a community centre or hotel presentation room, medical centre or school, microwave noise is substantial. Even at a town beach it may be significant. Visit dharamhouse.com for further information.

How do we respond to the onslaught of microwave radiation noise? What if we are playing cello in the symphony orchestra and suddenly the glorious sound is punctured by noise permeating each wall, the roof and floor of the concert hall?

1. Our own unique frequency on the cello (frequency of our cells and organs in our analogy) is disturbed. We lose our timing. We are unable to hear and follow the other instruments (heart, brain, pineal gland, pituitary gland and hypothalamus). During the Sydney EMF Experiment I found myself not feeling 'myself'. I did random things I'd never done before such as stepping off at Edgecliff Train Station instead of Kings Cross Station during a moment of brain fog. The instruments in my body–mind were struggling to hear their cue and follow.

2. We become fatigued with our role in the orchestra. We say, 'With all this background noise what is the use?' We throw down our instruments and stop playing. As we explore in more detail later cells shut down per the fight-flight stress response otherwise designed for when we are threatened by large animals. Many radiation pollution sickness symptoms may be attributed to this

shut down. Brain fog for instance is associated with neurological shut down. More sensitive instrumentalists such as the violinists (pineal gland) and flautists (pituitary gland) shut down intuitive connection even sooner.

One of the musicians deserves special mention. It is crucial in the quality of sound produced by the orchestra. The medical profession are aware of its importance and therefore in X-ray examinations the thyroid and parathyroid are protected by a thyroid collar. The thyroid is two lobes either side of the windpipe. An imbalance in thyroid secretions can contribute to anxiety and depression as well as menstrual irregularities and heart palpitations. Due to the superficial location of the thyroid and parathyroid these endocrine glands are highly susceptible to electromagnetic radiation penetration. Thyroid stimulating hormones produced by the pituitary gland increase with mobile phone use and there are possible deleterious effects of microwaves on hypothalamic-pituitary-thyroid axis which affects the levels of these hormones.[38]

It is not only the body–mind that is an electromagnetic symphony. What if we expand the symphony to encompass the entire human race and animal kingdom, Earth and the cosmos? Like our body–mind the universe oscillates in a magnificent symphony. It is similarly sensitive to layers of electromagnetic radiation.

THE OSCILLATING UNIVERSE AND THE HUM

My interest in earth vibrations intensified prior to the Sydney EMF Experiment from late 2012 when I heard The Hum. It was a low frequency vibration that only a fraction of the population hears. It sounded like a distant industrial generator drone of < 50 Hz. Science was and is baffled by it. There were attempts to link The Hum to geographical locations around the globe. Bondi Beach was a hotspot for instance.

This rumble followed me wherever I went. I was disturbed by the new lack of silence. I woke up one day to pour a morning cuppa and noticed a deep generator groan in my head. Earplugs did nothing to reduce the volume, indicating an internal vibration was taking place rather than sound vibration from an external source. I heard The Hum in various locations including a national park so deduced it was unrelated to industrial noise. Unlike tinnitus, it wasn't constant. The volume augmented at certain times of the day. The pre-sunrise and post-sunset periods are cherished in yogic traditions as ambrosial hours. These are times when divine connection is said to be amplified. The Hum volume magnified during these daily transitions.

I observed The Hum just as I significantly altered my way of being in the world. Beneath worldly success was held stress. During the period up to 2012 I held the light-heavyweight title for repression. I didn't want to explore the associated pain. Alongside the unwillingness was a deep desire to let go of what needed to be released. After a wrenching two years and the release of many layers of held stress The Hum volume turned down to almost zero.

Around the time of The Hum I was researching the apparent links between earthquakes and oil, gas, geothermal and water drilling. I was also receiving acupuncture from a friend to assist with lower back pain. Connecting the dots I asked myself, 'If pins on our meridian lines can ease aches and pains might energetic equivalents on the Earth's meridian lines or axiatonal lines ease its aches and pains also?' In years to come might holes be dug only in locations that balance earth meridians?

I'd often noticed the vibrational quality of land. The ancient wisdom transmitted by the land of the Pilbara and the Kimberley. The transformative energies of South East Queensland and Northern NSW. The vortexes of Sedona in the USA. Years ago I disembarked from trekking the depths of Tasmania's Tarkine Wilderness to feel the assault of a poisoned, clear-felled wasteland on its boundary. It was like walking belly-first into a bread knife.

A hole dug for coal extraction or a section of a land mass cut away to make space for a highway leaves a vibrational signature different to the surrounding pristine land. We are talking here about the intangible (or vibrational) value of old growth forests and clean rivers. Our body–mind is the greatest meter at present for measuring the vibrational signature. Upon setting foot in the Kimberley region of Western Australia we sense there is a special vibrational value in the land. Walking over a recently dug section of the Pacific Highway we notice the vibrational value has diminished. If science needs more than this then engineers are to develop a device that captures the oscillation rate of land and therefore its vibrational signature. A decision on whether to excavate or not could then be based on vibrational value.

Along with digging holes, removing trees and poisoning land areas we have telco and National Broadband Network (NBN) towers altering the vibrational signature of areas we live in.

On the island of Bali the phone towers are as ubiquitous as trees in a rainforest. Traditional Balinese elders confirmed what I felt – the energy of the island has changed dramatically. Hordes of tourists contribute psychic pollution. The bigger shift, they said, came from unprecedented tower construction and accompanying radiation pollution and disharmonious energies. Bali's symphony orchestra is out of sync and the music they produce is overlayed with noise. An orchestra can never sound sweet to the ear if hundreds of tall steel sentinels scrape the walls of the concert hall with irradiating fingernails. More on Bali and energy healers in later chapters. Back to Sydney and my dream-girl Katie.

CHAPTER 6

My Journey II – From Healthy to Hopeless

- *My ideal life starts to slide downhill*
- *The void in mainstream and alternative medicine*
- *The antenna across the road – living in a military-information age*
- *Creating a Zero EMF Sanctuary – and discerning the 'protection' product charlatans*
- *Realising that I had to leave Sydney*

I'D BEEN AWAY FOR WORK AND NOTED my malaise when flying back in to the city. It was radiation pollution sickness but I hadn't yet correlated the symptoms. I wanted to see Katie. She worked near the city. We met in Chinatown for lunch at a reputed noodle shop but I wasn't feeling the greatness. Tables and chairs were plastic and we were seated so close to another table full of students that I accidentally headbutted the chap behind me.

The noodle shop only did a monosodium glutamate-drenched version. I therefore opted for the only MSG-free dish available – a dull plate of Chinese 'vegetables' which was a singular green 'vegetable' drowned in oil and garlic. Four weeks earlier I might have enjoyed the outing and made light of my 'light' meal but that day my irritation levels were heightened. The microwaves had been accumulating and the morning flight home had taken me over the agitated edge. Why had the microwaves chosen me? I was jittery, hungry and dealing with a month-old 24/7 headache.

Katie devoured her noodles like someone who exercised fanatically in order to eat with fervour. We argued. She stopped eating and pushed her bowl to the side. I saw her turn into a leviathan. Chemical-drenched noodles and deafening Chinese student banter was the backdrop as our relationship took a slide. Things had been awry for weeks and we'd now hit a black diamond run without the relationship history or skills to handle the slope. For the first time I thought, 'Maybe I don't want to be with this woman.' I didn't usually quit so easily and felt tormented.

I projected an MSG monster onto Katie and later reflected I was an agitated electromagnetic radiation disturbed monster myself. It took very little for me to snap and I was so radiation pollution sick that I had no energy for a loving relationship. Stage 2 frustration (see Chapter 4) was in full force. My sense of intuition, though shaken by radiation pollution sickness, hinted that everything in my life was about to tumble out of control down that black diamond slope.

GETTING NOWHERE WITH HEALTH PRACTITIONERS

I felt like leaving the city for a few days to immerse myself in nature was a refreshing necessity. My symptoms diminished but on returning Monday morning they rapidly resumed. The confirmation from the weekend away was enough to dissolve any remnant denial. Technology was making me sick. I commenced correlating and scribed my observations. The Sydney EMF Experiment had begun!

I had hardly visited a doctor in previous years so it was a surprise to find myself in a waiting room for a medical check. I stretched the doctor's mind talking about 'toxic energies' and 'electromagnetic radiation'. When I enthusiastically pointed to my correlation with headaches, insomnia, anxiety, irritation and recent depression his eyes glazed over. He had no interest in investigating and just wanted to prescribe something. He ignored everything I said about microwave radiation but latched on to the recent depression and referred me to a psychologist.

If he'd known anything of potential adverse biological effects we would have had a better doctor–patient interaction. (Appendix B details BioInitiative Report 2012 biological effects data.) A handful of attributed biological effects at low intensity (everyday) exposures include –

- DNA damage – Our unique blueprint may be damaged and pre-cancer attributes formed.
- Blood-brain barrier opens – Allowing toxins into the brain and leading to neuronal damage. Occurs when phone at arms distance or at 1000 times lower than radiation exposure limits.
- Calcium efflux – Ca ions flow out from the brain attributed to altered neural function.
- Pineal gland interference – Melatonin production reduced (melatonin is a cancer protection) and serotonin production is altered leading to depression.
- Retarded learning and memory impairment in children.[1]

I made it to the psychologist but textbook psychology and talking about my childhood wasn't quite going to cut it. I got bored and started asking the psychologist questions. Soon we'd informally swapped roles. In spite of my guiding the psychologist towards life enhancement she billed me $150. I didn't return.

Doctors and psychologists have a critical role in today's world but my interactions with these professionals in searching for solutions were draining rather than fulfilling. I visited two more doctors and another psychologist then concluded that in the current medical paradigm –

- There was a knowledge void around electromagnetic radiation and its relationship to adverse physical and mental health effects.
- Disdain prevailed. The topic was not a familiar one and the professional would always redirect discussion back to their paradigm. Their understanding (current paradigm) was insufficient to deal with radiation pollution sickness. 'Take a painkiller for your headache.' 'Here is a sleeping pill prescription.'

This tendency to focus on what we know is limiting. For a GP medication and a follow-up visit is the solution. For a chiropractor manipulation of the spine is the solution and he tells the patient, 'I'd like to see you again next week.' It is one thing to hear those words after a date but hearing them repeatedly after visiting the chiropractor, psychologist, acupuncturist, kinesiologist or massage therapist means core issues are not being addressed. These may be useful pieces in a holistic approach but it is rare that a holistic approach is taken.

'I'd like to see you again next week' is a disempowering money-making crock-horse that the alternative medicine community rides as thoroughly as mainstream medicine. Has the healer worked his own way through whatever compromises our health? This includes held trauma, addiction, anxiety, insomnia and depression and radiation pollution sickness. Does she work with loving intent? After the session do we feel empowered to transform at home or are we relying on indefinite visits to feel okay? Does the healer walk the talk – has she created harmony and joy in her life or is her life one contracted event after another?

STRESS, CONTRACTION AND ELECTROMAGNETIC RADIATION AS A CATALYST

Telco antenna arrays, mobile devices and Wi-Fi toppled my sense of invincibility. During the Stage 3 correlation (see Chapter 4) I uncovered my kryptonite. There was an antenna less than 100 m away on a neighbouring apartment building rooftop. It was a small piece in the overall radiation exposure scenario, but my entire body tensed when I saw that antenna. I responded as if to a violent segment in a movie or eating a mouthful of food that I didn't enjoy. This occurred to a varying degree of subtlety with all electromagnetic radiation sources. On a macro scale my body would freeze and tighten up for a moment. On a micro scale my cells did the same in response to environmental stressors. During the Sydney EMF Experiment I was tense 24/7 at macro (body–mind) and micro (cellular) levels.

Very soon after moving to Sydney I attracted contracted events, frustrating relationships and material losses. Timing of life events during this period indicated electromagnetic radiation was a catalyst. There were other historical contributors including held stress and trauma but it was microwave exposure that tipped the scale and showed the most precise correlation with calamity. As long as I was exposed to microwave radiation the stress cycle perpetuated. Even when I later removed radiation by moving to a Zero EMF Sanctuary it took some time (and inner work) for residual physical and mental health symptoms and contracted life events to subside –

EMR > Cellular Stress > Physical and Mental Symptoms > Contracted Life Events

When a cell is subject to increases in environmental toxins, temperature or pH disruptions, there is a protective reaction called the cellular stress response. Pioneering scientists confirm the criticality of cellular stress in overall health and 95% of illness and disease may be stress related. Contrary to medical dogma our genes play a relatively minor role in health. In the cycle above the only way to improve physical and mental health is to reduce cellular stress and/or the extent and type of electromagnetic radiation exposure. We look at how to do this in later chapters but for now, while we are still in Sydney, we need a moment of Bondi Junction uber-cool.

During the Sydney EMF Experiment I went into an Apple store for the first time in my life. Except for an antique iPod Classic I'd never owned an Apple device and I'd never had urges to queue around the block for the latest iPhone. I had to see what all the fuss was about. If you are from Generation X or older then step inside for the culture shock. I went in clean shaven and with standard straight leg pants and came out with skinny jeans, short back and sides and a beard down to my nipples. It didn't help my radiation pollution sickness but people on the street said I was 30% cooler.

In 2013 Apple became the largest publically traded company in the world by market capitalisation.[2] This moment represented the final shift from military-industrial dominance to a military-information age. Bombs are passé. We are in the era of soft kill. Electromagnetic radiation is a gently undetectable way to influence and even reduce populations. Cyber-espionage, cyber-security and information warfare can all be used to destabilise.

After my puffed-up moment of uber-cool I resumed the slightly nerdy Sydney EMF Experiment and pulled out my radiation meters to measure antennas in Bondi Junction. Microwaves were relatively high in the area. With radiation pollution sickness exhaustion came on quickly so I only had a couple of productive hours in each day. I was relatively spritely at this stage. Later in the experiment some days I didn't make it out to buy food. After obtaining measurements I returned home to rooftop Potts Point where I was microwaved at levels equivalent to or higher than the Bondi Junction daytime exposure. It was impossible to recuperate while sleeping in that environment. I needed to build a Zero EMF Sanctuary.

ZERO EMF SANCTUARIES

A Zero EMF Sanctuary in my bedroom would make a difference. If I had seven hours of respite each day my jug of dark red raspberry cordial would at least partially purify with clear pure water. Perhaps my health would improve and I could continue living in Sydney?

The apartment's metal roof provided radiation reflectivity. Extensive glass doors and windows enabled full penetration of microwaves. Brick and timber clad walls did little to deflect radiation. Magnetic fields attributed to power lines were acceptable at 1 mG. Microwave radiation, however, was of extreme concern per the International Building Biology Guidelines in Appendix E. I measured 1000 $\mu W/m^2$ with amplitude spikes three times as high in the bedroom. These measurements are well within Australian radiation exposure limits and not uncommon in an inner city dwelling.

I flew in a shielding specialist, who suggested these options –

- Use shielding fabric for curtains to reduce the microwaves coming through the glass
- Paint the entire apartment with carbon (or graphite) based shielding paint plus shielding curtains
- Move out of the apartment

The only option financially viable was curtains. The shielding guy said this would reduce exposure levels by over half and possibly 75%. How did he calculate that? He didn't. He was selling a product. How would curtains covering the glass windows to the north reduce the radiation exposure if the walls to the south, east and west were being penetrated to a similar extent? I doubted the shielding chap's logic but I was radiation pollution sick and desperate for a solution. I accepted his 'expert' advice and paid for the curtains.

The day the curtains were to arrive I anticipated the possibility of a good night's sleep. I had some hope for recovery of what had been a thriving situation in Sydney. The curtains were couriered to my door and after I installed them it felt as if radiation levels had been turned from loud up to very loud. With radiation pollution sickness we can be caught in what I call the 'desperation hope devastation cycle'. We are desperate for a magic pill. Hope that it will work gives us some life force as we wait for the product in the mail. We are excited at the pinnacle of the cycle as we try it out. Hours or days later we realise the product was a farce and we've been duped. We slump into devastation that teeters on depression.

I slumped devastated on the microwaved couch then with my meters I analysed what had happened. A Faraday cage is a shielded space such that electromagnetic radiation cannot enter or leave it. My apartment had become a partial Faraday cage. The curtains prevented antenna microwaves entering through the glass doors to the north but radiation entered from the other directions and bounced off the curtains to amplify levels in some areas. There were pockets where readings were higher than without the curtains. I pulled down the curtains. I'd bought an ad-hoc 'solution' from a slick salesman. If I'd not been immersed in research and recently equipped with meters I might have installed the curtains only to cause more harm.

Companies popped up to sell grounding or 'earthing' products. I knew humans needed to ground to be coupled to earth frequencies and the Schumann Resonance frequency that oscillates between earth and the ionosphere. I knew how good it felt to kick off my shoes at the end of the day and stand on the grass. Linking grounding to electromagnetic radiation protection seemed a stretch however. I installed a grounding sheet on my bed and tried it out. My sleep quality while grounded for a week was no different to when ungrounded for a week. Electromagnetic radiation has a magnetic component. It pulses through our body whether we are grounded or not. When we are grounded our body becomes a conduit. Imagine our body–mind as a straw and a bottomless cup of raspberry cordial as electromagnetic radiation. A grounding sheet is a metal embedded fabric that connects our conductive skin to the earth via an electrical socket. The raspberry cordial still has to flow through our body–mind to reach earth. When it does connect to earth the same amount of raspberry cordial is in the straw and we are still exposed.

The desperation hope devastation cycle continued and despairing thoughts surfaced. I had to find a way to shake them off. The ocean had always been a magical healing potion. It was late winter chilly but I had an inspiration to swim. Woes would always disappear in the ocean. I'd always feel revitalised. I train and bus hopped to Bondi Beach. A handful of surfers braved the storm conditions. At the North Bondi water's edge I saw a blue-lipped chap in a wetsuit and a heavy ethnic lady frolicking as if she was immune to the cold. A couple walked along the beach hidden beneath hoodies. I felt the excitement of doing this thing I loved and ran out with calves flicking up like an ironman to plunge under a wave. Exhilarating. Yes! I body surfed a wave. Cold. One more wave. Yes! A little tube ride. Fun. Ye ah no. I walked up to shower realising my shivering skin was a momentary distraction but the jitteriness inside continued.

THE GOODBYE SYDNEY PARADOX

The kryptonite was too strong. After the Bondi episode my heart palpitations were more noticeable. I'd note a strong heart rate of 70 bpm which would randomly leap to 120 bpm and barely perceptibly weak. Four months into the Sydney EMF Experiment my drive to burrow deeper into the radiation pollution sickness rabbit hole waned. I was at a sufficiently dangerous bottom. What had once been stubborn fierce anger at being pushed from my Sydney dreams was now tired dejection. I was exhausted. The last weeks had been a blur and I was scared that I'd taken the experiment too far. Had headaches, jitteriness and brain fog and memory loss caused permanent neurological damage? Was depression going to accompany me from this point on?

Indicative of my disinterest in life I'd neglected to check my phone for days. Paradoxically it was a mobile phone call that triggered me to leave. I peered over the edge of the brick parapet wall the five storeys down to a festive café scene. On my rooftop oasis it was completely quiet. I let the Samsung smart phone ring four times before picking it up and gluing it to my ear. As I listened I saw mobile antennas within stone-throw distance set to the backdrop of the Westfield tower. Sage sticks, didgeridoo and the delicious laughter of the rooftop housewarming flashed through my mind. Late night strolls for an ice-cream elixir and sketching plans for an abundant rooftop food forest were fading memories. It was time to go. Later I would return to Sydney to share this knowledge and my path of restored health and spirit with others. The Sydney EMF Experiment was over. Suddenly it was urgent that I was gone from there.

During the Sydney EMF Experiment I observed that even after dropping off my mobile phone I was sickened by microwave radiation. The extent of passive electrosmog pollution, especially due to telco towers, meant even though I personally minimised my technology use I still had to leave Sydney. I think of telco and NBN towers now as steel sentinels spraying microwave radiation bullets 24/7. If we are in their direct line of fire the damage is severe. If we are outside of their trajectory there will still be collateral damage. Some sentinels are armed with machine guns. Others fire high-powered rockets and even nuclear missiles depending on the extent of 4G coverage the telco is seeking to achieve as is detailed in the next chapter.

CHAPTER 7

Antennas and the Radiation they Emit

- *The truth about telco tower emissions*
- *Tower trajectory and living in a giant microwave oven*

THERE WAS A TIME AT HIGH SCHOOL when I thought a girl named Rachel was lovely. I was too shy to talk to her. She was two grades younger than me so the only time I'd regularly see her was in line for canteen. Rachel was all I could see. I didn't mind if the lunch line snaked for a mile as long as she was in it. At seventeen the object of focus was motorbikes. For a period they were all I could see and I bought one to get me around in my first year of university. Once my awareness developed around electromagnetic radiation sources this focus went to telco antennas. As we drove along the beach one glorious day my friend commented on the shimmering ocean but I was busy observing the telcos antennas fixed to the town water storage supply.

In some towns antennas dominate the skyline. The Telstra land behind the post office usually contains a steel sentinel. In others they are tucked away behind pine trees or on a hill overlooking the main residential area. Sometimes they are disguised in rooftop facades, steeples, church crosses, fake palm trees, chimneys, flag poles, grain silos and water towers. Australian businesses are paid to conceal mobile towers and associated electrical stations from the public.

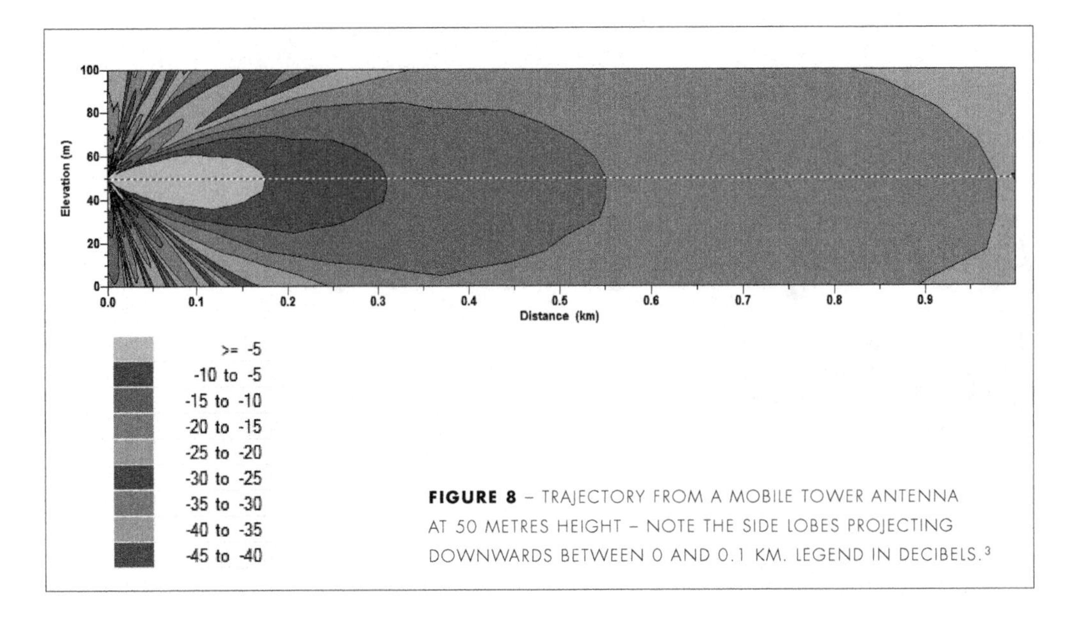

FIGURE 8 – TRAJECTORY FROM A MOBILE TOWER ANTENNA AT 50 METRES HEIGHT – NOTE THE SIDE LOBES PROJECTING DOWNWARDS BETWEEN 0 AND 0.1 KM. LEGEND IN DECIBELS.[3]

Figure 8 shows an example antenna trajectory. High levels of radiation exist beneath an antenna array from the side lobes projecting diagonally and downwards. Directly in front of the antenna remains the location of highest radiation exposure. As shown in Figure 8 the radiation intensity is especially high within 500 m. Distance is a critical factor according to the inverse square law detailed in Appendix D and Figure 15.

TURNING UP THE RADIATION VOLUME OF TELCO TOWERS

Just as we turn up the volume on our stereo to hear it further away telcos turn up power density of their antennas to enable their signal to be 'heard' at vast distances. An antenna array may have radiation outputs hundreds of times higher than a similar looking array. After my moment of cool after the Apple store visit I compared Bondi Junction antennas with antenna reference sheets found at http://www.rfnsa.com.au/nsa/index.cgi. When we type 'Bondi Junction' 14 sites are available. Site No – 2022002 at 1 Newland Street in Bondi Junction has an Environmental EME Report maximum reading 0.98% of ARPANSA exposure limits.[4] Readings are calculated at 1.5 m above ground level and are higher (data not provided) in the direct line of fire. Site No – 2022003 on the list is 130 Denison Street Bondi Junction. The EME report indicates a maximum exposure level of 10.33% of ARPANSA limits or power density 798.4 mW/m^2 at 1.5 m above ground level.[5] Again the readings will be higher in the direct antenna trajectory. This is 266,133 times the levels at which the BioInitiative Report 2012 says biological effects are observed.[6]

It outputs 798.4 times the level of extreme concern designated by International Building Biology Guidelines in Appendix E. Antennas in neighbourhoods can be found at http://maps.spench.net/rf/ or http://web.acma.gov.au/pls/radcom/site_proximity.main_page.

A while ago I attended a community discussion related to a proposed Telstra tower in a sleepy NSW beach suburb. According to those in the area phone reception was fine. Many were baffled as to why another tower was needed. They were Group 3 and seeking a healthier outcome for their suburb. Others were Group 2 ambivalent and didn't care if a tower was installed but NIMBY (not in my backyard). Preschool and public schools were < 500 m from the proposed location. While the proposed location was unacceptable to the majority the same participants were agreeable to Telstra building the tower elsewhere.

The community and local council was so used to being dominated by industry that compromise had become their default position. Even though phone service was adequate there was little energy for outright rejection of the tower. It was not understood that Telstra didn't care if they had to move. They might need to spend more on electricity due to the antenna output power increase in order to emit at desired levels. The initial proposed location was rejected. Community thinks they have had a 'win' but what will be Telstra's Plan B? Will they install a facility that exposes local children to higher radiation levels than in the initial proposal?

During the Sydney EMF Experiment my finances were calamitous even though I owned a rooftop apartment. In my financial predicament I could have contacted an Australian company to learn 'How to Market My Property for a Tower'. I lived in the rooftop fifth floor apartment adjacent to a taller building with Optus and Vodaphone antennas on its roof. The antennas microwaved my home at higher levels than the apartment building block on Optus and Vodaphone payrolls. The inhabitants of the building on telco payrolls were less likely to have radiation pollution sickness concerns. Does it seem fair? If I had a 300,000 lumen (blinding) tower spotlight that I shined through my neighbours windows all night would that be fair?

HOW MUCH RADIATION ARE TELCO TOWERS REALLY EMITTING?

I grew up eating fried lamb chops and sausages and oven-cooked chicken with potatoes. It was only when I moved into a share house at university that I discovered the benefits of a microwave oven. A diet of recyclable $1 Big Macs and slices of Pizza Hut meat lovers pizza sustained me for many years. Vegetables were so rare that I had to elevate my legs for shock after opening the fridge and observing a single carrot lying on the bottom shelf. After recovering I grated the carrot and added it to a beef mince bolognaise. Housemates were on the couch crying hunger. Due to lack of cooking pot options I used a large plastic bowl and slipped it into the microwave oven. Soon spaghetti bolognaise was ladled onto plates and we sat down to eat together. After two mouthfuls I noticed a blue plastic flavour amongst the tomatoes, garlic and herbs. I quickly cautioned the others and we went off for more salubrious options at Hungry Jacks.

I would not have been cooking with a microwave oven without Perry Spencer's work at Raytheon. This led to the first microwave oven just after World War II.[7] Raytheon Corporation was at the forefront of the military-industrial complex. They owned patents for and developed the Alaskan HAARP (High Frequency Active Auroral Research Program) facility in partnership with the US Air Force and British Aerospace Systems.

Microwaves are used in radar. Twenty years ago I voyaged up the west coast as a crew member on the sail training ship LEEUWIN. The captain warned us not to hang out in the direct trajectory of the rotating radar or else we wouldn't be able to have children. I wonder if the sterilisation warning is still provided given we now live in the trajectory of microwaves from tower antennas. Microwaves are also used to sterilise soil 'which is now accomplished with highly toxic chemicals'.[8]

Microwave oven frequencies are typically 2.45 GHz. The US Federal Communications Commission (FCC) also assigned 915 MHz, 5.8 GHz and 22.125 GHz. The microwaves emitted from telco towers are of similar frequencies as microwave ovens – the microwaves used to defrost a piece of steak or to reheat a lasagne are the same as the electromagnetic radiation pollution penetrating and delivering biological effects to our body–mind 24/7. Rather than being limited to four designated frequencies towers emit a vast array of microwave frequencies.

CHAPTER 8

Stealth Bombardment II – Learning from the Past

- *Looking at historical cycles and patterns to enable wise regulatory actions – cigarettes, asbestos, DDT and thalidomide*

WHY DO WE WAIT DECADES AFTER INITIAL health warnings before products are restricted or removed? Chernobyl occurred in 1986 and Fukushima followed in 2011. After Chernobyl nuclear reactors continued to proliferate. A few recent local examples –

- Thalidomide was produced by German-based Gruenenthal and sold in the 1950s and 60s as a cure for morning sickness. It was linked to birth defects and withdrawn in 1961. For every thalidomide baby that lived ten died. In 2012 the CEO of Gruenenthal issued its first apology.[9]

- DDT (dichloro-diphenyl-trichloroethane) was introduced in the 1950s as a pesticide. Reports and books warned of health effects of the chemical. The impact on birds was palpable. Despite this, in 1972 the Australian Academy of Science wrote a report 'The Use of DDT in Australia' recommending continued use of DDT. It was banned in the US that same year and completely banned in Australia 15 years later in 1987.

- ABC's FOUR CORNERS reported on 23 July 2013 – 'An urgent review is under-

> way after a FOUR CORNERS investigation found elevated levels of dangerous dioxins in a generic version of 2,4-D, one of Australia's most widely used herbicides. Dioxins are one of the most deadly chemical compounds in the world, but Australian authorities do not routinely test for them.' Dioxin contaminated chemicals affected Australians working in agriculture in the 70s and 80s – a legacy of 2,4, 5-T and 2,4-D herbicides which go into the Agent Orange mix.

The historical cycles for cigarettes, DDT, thalidomide, asbestos and now telco and mobile manufacturer microwave radiation emitting technology follows a pattern –

1. **Introduction** – The introduced chemical or technology is claimed to be all benefits and is swept into the market bereft of testing. The manufacturer may be aware of hazards (per mobile manufacturers and telcos) which are withheld from the public.

2. **Observations** – We notice a deterioration in the health of indicator species populations such as trees, birds, frogs or fish. Signs of adverse human health effects become evident.

3. **Evidence** – Clear links are made between the chemical or technology and toxicity and illness. The manufacturer seeks to prolong their sales through PR campaigns and creating confusion and doubt around the established evidence. We have had clear evidence linking biological effects of radiation exposure for over half a century.

4. **Action or Inaction** – An action would be reducing allowable levels or banning the technology. India responded to biological effects evidence and now have non-ionising radiation exposure limits 90% lower than Australia. They could go further but have at least taken a first step. Russian scientists showed biological effects half a century ago. They took action and Russia has exposure limits 100 times lower than Australia. Some choose inaction which leads to an unobstructed path for industry profiteering. Populations get sick and government is left to address residual medical and legal costs.

Australia can learn from history. We do not need to wait decades after health warnings have been issued as they have over and over with regard microwave radiation proliferation. We can surpass the actions of India, Russia and Switzerland by reducing radiation exposure limits by a factor 1000 by 2020. With a small population Australia has the relative nimbleness and flexibility yet we carry international weight. By prioritising health as the nation's number one value we can be at the forefront of change co-creation and inspire transformation globally. How?

Every decision made in business and politics would be checked – 'Is the health and harmonious development of the nation enhanced by this decision?' If the answer is no then it does not proceed. If the answer is yes then it is implemented. The media would do the checking and reporting so any decisions that perpetuate destruction are exposed. Suddenly the quality of the media information we are 'feeding' on improves. We would hear inspirational stories daily and there would be a societal shift from illness and disharmony to health and harmony. All from a simple question that all Australians are aware of and ask themselves daily – 'Is the health and harmonious development of the nation enhanced by this decision?' Yet we read this and we think 'yeah right', 'dream on' or 'too idealistic'. Why might we doubt our capacity to derail an old-paradigm steam train running on rusty tracks? Only because we've been trained to.

CHAPTER 9

Co-Creating Change

Why we accept telco and mobile manufacturer domination of the ether - and ways we can change the status quo

WE HAVE FORGOTTEN OUR TRUE NATURE AND our connection to Nature. Illustrative of this separatism is the way we label the 'environment', 'human nature' and 'nature' as distinct entities. We blame our 'human nature' for the exploitation of 'nature' as if we have no power to control ourselves. We dig another hole, build another Fukushima or construct another telco tower in the suburbs. 'We can't help it. It is human nature.' This is lazy thinking. It is not our nature to do any of these things. The false ego or fearful personality is the one exploiting. Why do we allow the exploitation? In my two years of research with electromagnetic radiation I've observed the enabling of unfettered telco antenna and mobile device growth is due to –

Denial – At school in the 1980s the worst thing a kid could be called was a 'dobber'. Speaking up was not encouraged. I saw classmates speak up and be so thoroughly rebuked by the teacher that I decided to play safe and keep my lips zipped. As we get older we carry the same fears of speaking out. A middle-aged dentist foregoes a paid workday to lock on to the bulldozer about to tear down the forest he has camped in annually since he was five months old. For expressing his truth he is ridiculed by mining executives and the media and even arrested. A woman is abused by a footballer and the entire football fraternity plus a multi-million dollar PR campaign turn on the woman to shame her speaking her truth. Bullies get their way.

We see the pattern of corporate domineering so why bother? We look the other way and hope things will resolve or someone else will step in to face the bullies. Knowing something needs to change but ignoring it is living incongruently. After a while this incongruence gnaws at our insides and we start to hurt. The path back to congruence is to stop denying and take action. We are not alone with 3-5% of Europeans experiencing radiation pollution sickness.[10] Magda Havas is a leading researcher who believes 1-5% of western populations have extreme symptoms and 35% of the population have mild to moderate symptoms.[11]

SOLUTION – SEEK HARMONY BY SPEAKING OUR TRUTH AND TAKING ACTION.

Seeds of doubt – Mobile and telecommunications industry 'experts' and their PR machine sow seeds of doubt. They generate doubt as to the science behind adverse biological effects of electromagnetic radiation. This strategy has been used by propagandists for decades. The invisibility of electromagnetic radiation and current lack of knowledge surrounding it results in a public easily led by the PR feed.

SOLUTION – EMPOWER OURSELVES THROUGH QUALITY KNOWLEDGE SO WE ARE ABLE TO SEE BEYOND THE LIES.

Self-accountability – Industry forces guide technology but we hold accountability for driving technology sales. When we stop buying the latest iPhone we stop feeding industry.

SOLUTION – SEND A MESSAGE TO INDUSTRY WITH A DECISION NOT TO PURCHASE.

Device addiction – For years I needed to be available around the clock for my work. I used my mobile in the day then placed it on the bedside table when I slept. Rarely was it more than a metre away. On occasions I was one of 'those' guys. The one at a restaurant conversing with friends when a loud beep emanates from his pocket. He halts the discussion to look at a flashing phone and say, 'I need to take this.' He answers and walks outside the restaurant pacing figure eights. Our eyes follow him. Al fresco diners note his confident swagger. Some think, 'He must be an important gentleman stepping outside to answer.' His ego gets a little puff as he returns to the table. He asks, 'Where were we?' as if now that he is back at the table the conversation can recommence.

I worked with a few smokers over the years and was curious as to why they did it. They smoked to relax. They would usually go outside for a puff with friends thereby meeting their need for connection. With the long exhale smoking slowed down the breath and got people present. For some it was a meditation and escape from anxiety.

Mobile devices are another mechanism to escape from anxiety and overwhelm. Paradoxically they contribute to our anxiety and irritability through microwave induced cellular stress then seduce us as being a way to feel better. We scroll the hypnotic screen's social media, YouTube videos or news bulletin seeking comfort.

SOLUTION – BE BRAVE AND SWITCH OFF WITH OTHERS FOR A DAY, A WEEK, A MONTH AND CONTRIBUTE TO TRANSFORMING SOCIETY AT MOBILEFREEDAY.ORG

Lack of creativity – My first job out of university was as a design engineer. One of the initial lessons I absorbed was that restraints inspired the creative process. With plenty of budget and no technical limitations the tendency was to design 'business as usual.' We did things in the safe, dull way that had been done before. When the budget was reduced by 75% and environmental impact was to be zero our team had breakthroughs. Suddenly we had to create a new paradigm. Since then the word 'restraint' has become one of my favourites. It equals creativity.

What if NBN was required to run 100% fibre optic rather than fixed wireless towers? What if schools and hospitals were deemed Wi-Fi Free Zones from 2017? What if engineers were required to design technology around radiation exposure limits 100 times lower than they are today by 2017? What if pulsed modulated waveforms were not permitted to be transmitted? What if we implemented radiation exposure limits 1000 times lower than today by 2020? What if we decided to reduce radiation exposure limits by 1,000,000 times by 2030?

SOLUTION – USE HEALTH ENHANCING LIMITS (OR BOUNDARIES) IN OUR OWN LIVES TO SEE WHAT IS CREATIVELY POSSIBLE.

CHAPTER 10

My Journey III – Leaving for the Country in a Low EMR Car

- *Leaving the city in style in my 'low radiation' Merc*
- *Becoming a radiation refugee*

MOVING INTO THE SYDNEY APARTMENT WAS LIKE diving head first into a pot of boiling radiation soup. The acute onset of symptoms made it easier for me to correlate and take action. Along the journey I've met many who experience mild to moderate symptoms. Their discomfort is not as substantial so their resolve to find out what is going on may not be as ardent. Sometimes they take no action. Sometimes they go from one practitioner to another with little improvement.

Caitlin came to see me after doctors kept telling her she was fine and she'd spent three years on the merry-go-round of alternative therapies. She said, 'Nothing seems to work. I am tired and irritable and my headaches won't budge. Otherwise I lead a functional life with a job and a daughter. Sometimes I wonder if I'm a hypochondri-ac.' I asked about her internal condition and her external circumstances. It was clear both could do with a shift as they always can. She'd been working with kinesiology to change her beliefs. After a few days the beliefs reverted back to limiting. There were sabotages, suppressions and core issues unaddressed. I use kinesiology with clients in combination with other methods so I'd seen this previously.

Caitlin said she had wireless in her house running 24/7. She occasionally took her computer to bed to watch a streamed movie. Her phone went with her wherever she went. We trialled switching off technology. This helped with her irritability and reduced the intensity of her headaches. Collapsing held stress and trauma completely shifted them. A side effect was that when we transformed her limiting beliefs the new empowering beliefs stuck. Prior to our first meeting Caitlin had considered her diet and chemicals in her beauty products but had not considered electromagnetic radiation as a potential trigger and root cause. She'd thought about her current family life and associated stresses but this was symptomatic stress compared to root cause 'forgotten' traumas from childhood and even earlier. In three sessions Caitlin reconnected with her radiance and was empowered with her own methodology to 'tune up' at home.

Before I could work with clients like Caitlin I had to get moving out of Sydney. Flying was out of the question due to the severity of my symptoms. I decided to buy a low radiation vehicle for the journey.

OH LORD, WON'T YOU BUY ME A MERCEDES BENZ

I was experiencing extreme radiation pollution sickness at that time so every choice was overlayed by 'Will it exacerbate my symptoms?' When I was buying a car I was limited in choice. Cars are partial Faraday cages with the metallic shell reflecting emissions that are generated inside such as those from a mobile phone. The shell also reflects emissions from towers. I've found a low radiation vehicle preferable to flying and train travel. But which car does an electromagnetic sensitive individual or someone with concerns for their small children buy? Many new vehicles have embedded microwave emitting electronics while some older vehicles I tested had high levels of ELF-EMF (magnetic fields).

The only way is to know is to measure. I'd like to see *EMR Exposure Stickers* on new vehicles sold in Australia by 2017. Stickers would provide information on microwave radiation, magnetic fields and electrical field levels in each seat of the vehicle. Back seats furthest from the engine and electronics are most suitable for children and electromagnetic sensitive adults to ride in.

I was so disturbed by radiation and financially wracked at this point that I had two purchase requirements – low radiation and less than $3000.

An old Mercedes in Sydney's south matched my requirements. The Mercedes had an original cassette player but I forgot to bring my 'Hits from the 80s' cassette tape. It was a summer-hot day in spring as I set off on the highway out of Sydney. I wound down the windows and turned up the fan. Wisps of smoke lifted from road embankments that had been bushfire-scorched the day prior. There was a burning sensation in my feet and lower legs emanating from the pedals. Was it from the heat of the bushfire? The leg burn was perplexing and soon unbearable. I pulled over and found a 5-foot stick which I jammed against the accelerator. This enabled withdrawal of my feet. At 90 km/hr I propped the stick against the armrest for makeshift cruise control. I was free to sit in lotus pose or stretch my legs above the dash, instigating raised eyebrows of passengers in overtaking vehicles.

The throb in my right foot continued and I stopped once more. The car had been purchased because of the low levels of electromagnetic radiation confirmed at the test drive. What had changed? I retrieved my meter and started the engine. Magnetic field readings were 5–6 mG at my torso, which is of extreme concern per International Building Biology Guidelines (see Appendix E). The needle maxed out near my feet at the pedals. Readings were hundreds of mG at the accelerator pedal foot and therefore hundreds of times the level of extreme concern. I switched the fan off and readings went to around 1 mG per the test drive.

I'd gone from being microwaved in my new apartment to electromagnetically magnetised in my new car. The next morning my toes were sore and there was a ringing in my kneecaps. After three nervous months of sore, tingling limbs the magnetisation symptoms subsided. I kept the retro Mercedes fan in OFF position. When I parked around town kids that looked like I did the moment I left the Apple store asked me about the vehicle. I'd bought it purely for health and financial reasons but I was an early unintended pioneer of the retro vehicle revival in my area.

Unfortunately for the hipsters retro does not always mean less electromagnetic radiation. I picked up a $30 second-hand Sony Trinitron television four years ago when everyone else was going digital. It had a small screen around 30 x 40 cm so my options were to use binoculars or sit up close. At the screen magnetic field levels were hundreds of mG so of extreme concern and on par with the Mercedes fan.

Magnetic field strength declines rapidly with distance. At 4 m away the field contribution was negligible. A modern LCD TV is preferable in terms of radiation levels. Sometimes (like the fan in the Mercedes) it is deceiving. The appliance looks like it would be low radiation but is the opposite. The only way to be sure is to measure.

RADIATION REFUGEES

Almost 90% of Australians are urban dwellers. After the Sydney EMF Experiment living in a city or town was not possible for me. I remember the sense of unfairness when I realised I could no longer enjoy the perks of the city – the high paying jobs, the food, people and city buzz. My freedom felt severely restricted. At the same time my interest in these things had dwindled in recent years. I'd always had a foot in the countryside and I'd find the funds I needed. I was fortunate to have been so indifferent to city life.

But many experiencing radiation pollution sickness are leaving a lot more behind. Imagine the impact of relocation on a family with kids at a prestigious school or simply someone who loves city life. What really hurt me were the financial impacts of being in refuge. I was not in a state fit for work for many months after the Sydney EMF Experiment. When things became urgent there weren't many work options in the area I took asylum in. At one point in refuge I had to sell everything I owned including a treasured stack of surfboards.

Once I accepted moving to a remote location for a period was part of my Path I no longer felt like a radiation refugee. I no longer felt like I was missing out on life. I transformed my definition of freedom to one of being okay with where I was at no matter what the outer condition. There may have been other ways to maintain a city life if my symptoms were less severe. I could have left the city every weekend for recuperation. I could have spent money to shield, though this has drawbacks including altering subtle frequencies required for thriving health. My symptoms were acute and I had to go remote. If the move beckons you can ask –

- Could this be part of my Path?
- What have I dreamt of doing in a farmhouse in the middle of nowhere? For example, painting canvasses, learning a language, growing organic vegetables.

- Is this a hidden blessing for my family and young kids? The prestigious city school was over-rated and attracted a cold bunch of humans anyway.

- Will this move create space for healing internal trauma and stress as my journey requires? I have not had the time to go deeper such is the fast wheel of city life.

- Is it a good time to sell our home adjacent to an antenna tower before microwave antenna towers really start to impact property prices?

CHAPTER 11

Wi-Fi Strong-Armed by the Telstra Mob

- *Discovering that Telstra can remotely switch on wireless – even when you've set it to OFF*
- *Observing that the only way to have Telstra turn off wireless is via legal means*

THE MERCEDES ROLLED INTO THE ZERO EMF SANCTUARY driveway. I was there to demagnetise and radiation detox in a rural retreat. With my new meters I was baffled to find high frequency electromagnetic radiation levels at the couch similar to those in Sydney (> 1000 $\mu W/m^2$ of extreme concern per Appendix E). The modem had been tucked underneath the couch for visual amenity. How could there be such high microwave exposure when I had hardwired via Ethernet cable a year prior to the Sydney EMF Experiment? The 'Wireless' light on the Telstra provided Technicolor TG587n V3 modem was off. There was a hardwire cable back to my computer with sockets flashing. So the Wi-Fi was off right? Nope. Further investigation confirmed there was no way to physically switch off the wireless. Even though the 'Wireless' light was green when ON and there was no light when OFF the modem continued emitting microwaves when OFF.

If I'd not spent thousands of dollars and hours on research and equipment I may not have discovered this issue of covert exposure and it may have gone on indefinitely. The discovery was disturbing. Generally I'd switch the internet on in the morning and off just before bed. I'd been diligent enough to hardwire yet my head had been 20 centimetres from a wireless emitting modem as I read a book on my couch. This situation led to cumulative electromagnetic radiation exposure. Constant increments of raspberry cordial were added to my jug of water each day I switched on the internet and I was therefore more susceptible to rapid decline once I moved to Sydney.

How did I resolve the Telstra strong-arm issue? This phase of the journey was an introduction to the extent telco tentacles influence our lives. Telstra can remotely switch on wireless whenever they wish.

After multiple calls to Telstra I established the way to disable wireless. One day I had internet speed issues and after I called them and reset the modem the settings went back to wireless enabled. I again disabled wireless using the steps below but Telstra kept remotely resetting my system to automatic wireless emitting. I had to check my modem every hour because Telstra were persistent in their efforts to have me using wireless. A blue light indicated no wireless then I'd go and make breakfast and the light turned green and wireless was emitting. Telstra did their very best to keep my Wireless mode ON and calling them was a futile exercise. I detailed a complaint letter claiming precise damages. The day the complaint letter was received things changed. The light stayed blue and wireless remained off. Telstra stopped tinkering with my system. Since that letter my modem has not once switched back to wireless emitting. Here's how to disable wireless on a Bigpond Technicolor TG587n V3 modem –

1 Go to 10.0.0.138
2 Technicolor Gateway
3 Home network
4 Interfaces
5 Select WLAN/Wireless Access Point
6 Configuration
7 Pick a task … Configure the main WLAN
8 Unclick 'Interface Enabled'
9 Check modem light goes from green to blue

You are wireless free and your hardwire set-up is actually being used for the reason it was installed! After the insanity of this episode I asked myself, 'How does it feel to live in a society where a Goliath like Telstra can flex their muscles to remotely irradiate me or friends or family without repercussions?' 'What is it like to request action from Telstra a handful of times and only receive respite when I send a legal drawn letter?' It is as disturbing as the microwave radiation waveforms they emit from home modems and neighbourhood antennas. It is simultaneously an inspiration to 'reset' and rewire society. But first a dhal curry with friends.

CHAPTER 12

Electromagnetic Radiation Exposure Limits

- *How microwaves are cooking our organs and brains*
- *Thermal effects versus biological effects*
- *Public exposure limits are millions of times higher than levels of observed effects on humans*
- *An electrosmog future as new technologies are released*
- *Microwave radiation parallels with nuclear radiation*

SOME YEARS AGO I HAD TEN HUNGRY people in my house with two gas cooktops. It was a juggle to time all the dishes to be sizzling without a reheat. The sweet potato chips were slow to cook so I fired up the gas to maximum. A dhal curry was ready simmering. I was bothered by the basmati rice lying idle in a pot and getting the early winter shivers. The crisp-skinned chips charcoaled after a single minute distraction to pour a glass of wine for a guest. Smoke filled the kitchen and they were removed from the menu. When proteins are heated to the point of browning heterocyclic amines are present, some of which are carcinogenic.[1] The foods we eat become carcinogenic with excessive heat exposure. What happens to our organs when cooked?

If the sweet potato chips had been broccoli they would have burnt even sooner. If they'd been lamb chops they'd have taken longer to burn. Similarly our skull and internal tissues are made up of materials and fluids of different density and conductivity. The tissue has varying susceptibility to burning from microwave radiation. Ridges, prominences and voids, variable head size and structure and non-uniformities mean the radiation level that burns us may not burn another person and vice versa. A 1955 study showed the danger of burns when radiation exposure was over bony prominences, primarily in the brain with frequencies above 750 MHz.[2] Mobile phones happen to operate in the vicinity of 750 MHz and above. Many of us have already cooked brain tissue and destroyed micro-areas of our brain.

In 1977 researchers Lin, Guy and Caldwell proposed that even at low absorption levels microscopic hot spot destruction may be occurring unnoticed. They showed 0.1 mW/cm^2 average power density led to 140 mW/g Specific Absorption Rate (SAR) 'hot spots' in irradiated animals. They had earlier determined that average power density 0.1 mW/ cm^2 should result in a SAR of 0.09 mW/g.[3] Therefore hot spots of greater than 1500 times the expected level were obtained. The corollary is that small areas inside our skull may be burnt and destroyed at levels below manufacturer required mobile phone SAR ratings. After 20 minutes talking on a mobile phone we feel the burning sensation at our ear but hot spots have already caused brain damage. The military-derived thermal limits did not prevent the cooking of micro-areas of our brain.

OUT-OF-DATE THERMAL LIMITS

A decade ago I was in Los Angeles for a Tony Robbins seminar. We entered an energised state and while reciting an affirmative mantra walked across burning coals. Afterwards the soles of our feet were unscathed. Prior to the three-second walk we prepared for hours and recited a mantra to get into a semi-hypnotic state. I haven't met anyone who prepares for three hours and uses affirmations before they subject themselves to a brain-burning mobile phone call.

Australia followed the international guidelines set by the World Health Organization inspired ICNIRP to use thermal effect based exposure limits. In Australia unless radiation cooks our flesh it is permitted. As we have seen above this is dubious science given hot spot absorption. The other dubious aspect is SAR averaging time. SAR is the mass normalised rate at which radiation is coupled to biological tissue in watts per kilogram. The unit 'watts' is indicative that SAR is a measurement of rate of heat transfer to tissue. In Australia we say 2.0 watts per kilogram measured over 10 grams of tissue is safe. SAR averaging time is based on a 6-minute call period. If we speak for over 6 minutes on our mobile phone we have likely exceeded the thermal measurement criteria.

The standard setting 'expert committees' have not kept up with technology proliferation. We are exposed to microwave levels often millions of times what we were exposed to two decades ago. However, non-ionising electromagnetic radiation limits remain thermal effect based, established during a military era when microwave towers, wireless cafés and mobile devices did not exist. The closest thing they had to what we have today was military radar which was used in war zones – and personnel knew to stay out of the trajectory. Substantial revision of limits has not occurred to incorporate biological effects (or biomutation) evidenced at low level exposure.

BIOLOGICAL EFFECTS OF MICROWAVE RADIATION ON CELLS

With leaky gloves and boots I skied sub-freezing conditions at Lake Louise in Canada. At the end of the first day I crouched beside the après ski bonfire with flames licking my semi-frost bitten extremities. A rugged old skier saw where I was and vigorously suggested I get back from the fire. He pointed out that I was so numb I didn't notice the cellular stress from the heat. He reminded me it was a shock for our bodies to go from hypothermic to flame grilled.

Our cells react to radiation long before physical burns occur. The BioInitiative Report 2012 says 'the body of evidence at hand suggests that bioeffects and health impacts can and do occur at exquisitely low exposure levels – levels that can be thousands of times below public safety limits'.[4] It says, 'Non-thermal effects of microwaves depend on a variety of biological and physical parameters that should be taken into account in setting the safety standards. Emerging evidence suggests that the SAR concept, which has been widely adopted for safety standards, is not useful alone for the evaluation of health risks from non-thermal microwave of mobile communication. Other parameters of exposure, such as frequency, modulation, duration, and dose should be taken into account. Lower intensities are not always less harmful; they may be more harmful. Intensity windows exist, where bioeffects are much more powerful.'[5]

The Australian General Public EMF exposure limit for a common 2.6 GHz telco signal is 10 W/m^2 or 10,000 mW/m^2 or 10,000,000 $\mu W/m^2$ or 1000 $\mu W/cm^2$. The limit varies depending on emitted frequency.[6] Byron Bay's green camouflaged antenna array in the heart of town at 50 to 100 m distance has an existing and proposed equipment power density level of 326.51 mW/m^2, which is 4.77% of ARPANSA exposure limits.[7] This value is measured at 1.5 m above ground level. In the line of fire it is considerably more as I discovered when I climbed an embankment to take close up (100 m) photos. Sixty seconds in the beam was a reminder that even though I am now far less affected than 12 months ago, minimising exposure remains good practice. At 50 to 100 m and 1.5 m from ground level it emits radiation at 108,836 times the levels at which the Bioinitiative Report 2012 concludes biological effects are observed. Exposure levels are 326.5 times the level of extreme concern designated by International Building Biology Guidelines (see Appendix E).

Another Vodaphone, Optus and Telstra array on the town's water storage supply has an existing and proposed equipment power density level 38.09% of ARPANSA exposure limits at 50 to 100 m distance.[8] This is 2460.36 times the level of extreme concern designated by International Building Biology Guidelines (see Appendix E). When emissions from the other antennas are stacked radiation exposure in parts of Byron Bay town are not dissimilar from those in a capital city. This is not atypical for a regional town or city.

The BioInitiative Report 2012 concludes biological effects are observed at 0.3 nanowatts per square centimetre or 0.0003 $\mu W/cm^2$. They suggest 0.0003 $\mu W/cm^2$ as a precautionary action level for exposure to pulsed radiofrequency radiation. The Australian General Public exposure limits are at a factor of over 3.3 million times higher than the levels of observed biological effects and recommended exposure levels.[9]

We might ask the question 'Am I being exposed to anything near Australian General Public exposure limits?' The answer is – absolutely. In some cases the limits are exceeded due to the radiation interference of multiple sources as described in Part I. Bondi Junction had 14 antenna sites at the time of writing with most of those sites catering to multiple telcos and therefore multiple frequencies. Telco 'gods' are now firmly established on every suburban block. They dominate the electromagnetic spectrum by 'owning' the universal ether or fluid that connects all things. Telcos control radiation transmission power, frequencies, modulation and trajectory.

LOOKING INTO A 5G FUTURE

During high school geography class I was introduced to the tragedy of the commons. We discussed how self-interest of the few could damage the livelihood of many. The 'commons' are land and resources belonging to and affecting the entire community. For the benefit of the community these resources were not privately owned. Until now. In Australia and internationally the sky and ether has been sold off to telcos. The 2013 Digital Dividend auction led to a government sell-off of Australia's remaining spectrum. Telcos obtained 15-year licences to expand their operations into strategic blocks including 700 MHz, 900 MHz and 2600 MHz bands. 700 MHz provides radiation penetration and 2600 MHz provides high data rates. These bands are additional to existing frequency bands so they are additional layers of radiation for our body–mind to process.

Long Term Evolution (LTE) communication technology (or 4G) was introduced in 2009 without studies. The first study on the short-term effects of LTE using MRI scans found mobile phone radiation with a phone 1 cm from the participant's head for 30 minutes altered brain activity on both sides of the brain. This was not only the near region but also the 'remote' area of the brain.[10] Brain alteration will be more substantial if the phone is pressed to a user's ear.

5G network availability is likely near 2020 with average speeds on networks of 10 Gb/sec up to 800 Gb/sec.[11] This will benefit anyone excited to stream ten movies on their device simultaneously. For the rest of us it is another unexamined technology that would stack onto existing 4G and 3G antennas and existing microwave radiation electrosmog. Those extra gigabytes can only come from a higher powered signal traversing the telco owned spectrum to reach our device. Our health will be collaterally damaged whether we adopt 5G, stay with 4G or decide to switch off. The steel sentinel's invisible bullets continue penetrating.

There are parallels in adverse health effects attributed to nuclear radiation and electromagnetic radiation. Both are invisible and can debilitate a chimpanzee or human within seconds. What if we viewed steel sentinel telco towers the same way we view nuclear reactors? Or if ARPANSA regulated microwave radiation sources such as antenna towers as thoroughly as they do nuclear sources? What if the similarities of effects including DNA damage and cancer risk were shared with the public?

THE NUCLEAR–ELECTROMAGNETIC RADIATION CONNECTION

In March 2011 Fukushima was the site of the largest nuclear disaster ever. There was an information blackout. Media coverage of cesium-134 and -137 being dispersed via ocean and air currents and impacting Pacific fisheries and the west coast of the Americas was negligible. A senior scientist at Woods Hole Oceanographic Institution analysed thousands of fish and found high levels of cesium-134 that could only be coming from Fukushima. Ken Buesseler said 'It's getting into the ocean, no doubt about it......The only news was that they finally admitted to this.'[12] A researcher of Fukushima talked about a million dying and eight million 'crippled' in the 25 years post-Chernobyl. Due to Fukushima's radiation release being around six times that of Chernobyl he predicted in the 25 years post-Fukushima there would be seven million dead and fifty six million 'crippled'.[13]

Evidence collected from as far back as the World War II Hiroshima bombings shows that ionising radiation has a 'leukaemogenic effect on man' beyond reasonable doubt.[14] The KiKK study looked at all cancers at the 16 nuclear reactor locations in Germany from 1980 to 2003 and linked leukaemia in young children living within 5 km of a reactor.[15] In 2002 the International Agency for Research on Cancer (IARC) classified extremely low frequency magnetic fields as possibly carcinogenic to humans. This was based on a doubled risk for childhood leukaemia at 0.3–0.4 mG exposure.[16] This level is exceeded in most homes.

The US National Academy of Sciences confirmed that 'the current scientific evidence is consistent with the hypothesis that there is a linear, no-threshold dose-response relationship between exposure to ionizing radiation and the development of cancer in humans'.[17] It is saying that even the most minor exposure to nuclear radiation causes biological effects which can lead to cancer.

Blank and Goodman showed in 1994 that cells required a threshold energy of electromagnetic radiation stimulus over 1 billion times weaker than a thermal stimulus for a cellular stress response. Blank observed that because 'cellular stress response is a reaction to potentially harmful stimuli in the environment, the cells were asserting that EMF is potentially harmful to cells'.[18] As we have seen, exposure limits are currently thermal-derived. Based on the Blank and Goodman study to avoid the electromagnetic radiation (technology) induced cellular stress response our exposure limits would need to be set 1 billion times lower than is current.

The BioInitiative Report 2012 indicates 'Cancer risk is related to DNA damage, which alters the genetic blueprint for growth and development. If DNA is damaged there is a risk that these damaged cells will not die. Instead they will continue to reproduce themselves with damaged DNA, and this is one necessary pre-condition for cancer.'[19] Lai says the genotoxic 'effects of both RF and ELF fields are very similar'.[20] In other words the microwave radiation from Wi-Fi and mobile devices is similarly genotoxic to radiation from fields elsewhere on the spectrum including extremely low frequency high voltage power lines. Electromagnetic radiation has been confirmed genotoxic just as nuclear radiation has been confirmed genotoxic. Appendix F is a summary by Neil Cherry of observed effects including genotoxicity.

A BRIEF HISTORY OF EXPOSURE STANDARDS

After World War II radar system power outputs increased significantly. In 1951 a microwave technician for Sandia Corporation complained of blurred vision and was found by the Sandia medical director to have cataracts and acute inflammation of the retina. The technician had been exposing his head by checking the antenna. Physicians, microwave operators and ophthalmologists were alerted through the case to 'the potentialities of microwave radiations in order that the use of this form of energy will be accompanied by appropriate respect and precautions'.[21]

Microwave exposure limits evolved from US Navy set 100 mW/cm^2. In 1966 they were reduced to 10 mW/cm^2 for the first microwave standard.[22] The Soviets reported biological effects at levels below 10 mW/cm^2. They set their occupational exposure limits one thousand times lower and public exposure limits ten thousand times lower. They considered non-thermal effects of cumulative exposure on reproductive and genetic health.[23]

In 1967 US President Johnson asked Soviet Prime Minister Alexei Kosygin to cease assaulting the US Embassy in Moscow with microwaves. In 1962 the US had discovered that microwaves were being beamed into the embassy at exposure levels 500 times lower than US worker limits.[24] The US approach was to check the health of the embassy staff while telling them they were being tested regularly because of a 'Moscow viral infection'. The staff were not advised they were being irradiated.[25] Health studies carried out on Moscow Embassy employees was reviewed by epidemiologist John Goldsmith. Evidence indicated chromosomal changes, haematological changes, reproductive effects and increased cancer incidence from the microwave irradiation.[26]

Australia informally followed the US thermal effect based standard 10 mW/cm². In 1979 David Hollway from the CSIRO formed a committee to draft an Australian Standard. Via the TE/7 Committee Australia had its first standard in 1985.[27] Hollway had read (or was aware of) the Soviet literature on biological effects and initially proposed 40 µW/cm². Industry and military interests didn't accept his proposal and 200 µW/cm² for public and 1000 µW/cm² for workers was adopted. This was significantly higher than Soviet exposure limits.[28]

Hollway stated in 1985 that the procedure of setting permitted levels at not too large a factor of safety below danger levels was appealing 'to those owning or controlling sources of radiation' and that the 'procedure is unintelligent at best…'[29] In the 1980s there weren't microwave emitting antennas on every block and potential disruption to the health of 23 million Australians as today. Despite the burgeoning exposure recent decades are devoid of actions to enhance public health. Limits are currently 1000 µW/cm² or 1 mW/cm² for the public exposed to a 2 GHz microwave source.

Changing the regulatory landscape to be one supporting health and harmony rather than an industry-dominant money control paradigm may seem daunting. There are examples of groups who came together for the values of health and harmony – and sustained. The Waterfall community south of Sydney is one that came together in the early days of microwave towers.

In 1995 the Waterfall community in New South Wales voiced their concerns over the pending construction of a mobile base station next to the local school. At a meeting with Telecom (now Telstra) scientists and officials, a Telecom representative said, 'How did these people get to know so much?' Public pressure had Telecom dismantle and remove the base station. The informed folk in Waterfall rose up to co-create a healthy school environment and the Telecom Goliath relocated the base station for 'technical reasons'.[30]

'How did these people get to know so much?' With the knowledge shared in **Playing GOD**, Dharam House EMF Essentials talks (dharamhouse.com) and online sources we can all become thorns in the side of telco corporates. We ask an informed question and suddenly the public relations representative or corporate 'expert' is stumped. He realises that 'these people aren't as clueless as I expected them to be' and plans are backed off. He goes back to the office and suggests to the division manager, 'The people in this community are too sharp. They know all about radiation and adverse health effects and will not allow a tower in. My recommendation is we pull out for technical reasons.' Imagine if all around Australia telcos and NBN were dismantling and cancelling installations for 'technical reasons'.

CHAPTER 13

Some Legal and Financial Aspects of Irradiating Your Neighbours

- *Radiation may soon impact neighbourhood relations*
- *Legal cases in the US and Australia*
- *How towers and antennas affect property prices*

I ARRIVED AT A PROPERTY I WAS to live in for the next year to see a disaster zone straight from APOCALYPSE NOW. There were piles of debris 2 m high and holes in the walls big enough to allow in the wildlife. I was there because it had 'potential'. After a week-long clean-up the functional and celebratory bonfire was of Burning Man proportions. Flames scaled the height of large forest trees. The fire burned all day and settled into a smoky smoulder. In the afternoon the wind perked up unexpectedly and blew smoke directly north east onto the neighbours Hills Hoist-hung washing. Knickers, socks and Saturday night dresses had their fresh washed lavender scent extinguished by billows of smoke.

When the neighbour complained I proposed it to be a perfect day for smoking salmon on the Hills Hoist. A laugh was not elicited. Smoke is difficult to block and redirect. It traversed their closed doors and windows. I offered to wash their clothes and carpet and confirmed it was a one-off event and I would not be burning off again while I stayed there. Neighbourhood relations remained frosty from that day on.

This was an accident. What if our neighbours smoked us out 24/7? We'd close doors and windows and seal them with tape. We'd ask the neighbour to urgently dowse their fire. We'd call regulatory bodies if no action was taken. With a young child in the mix we might urgently relocate to avoid damage to their sensitive lungs.

We can imagine the electromagnetic radiation equivalent as black smoke that travels through walls, windows and in the case of power lines penetrates the earth. Plumes of black surround the power lines hanging or buried at the front of our house. Black smoke emits from the mobile phone on our kitchen table. There is a black cloud around our home entertainment centre. The wireless modem in our study makes the room so dark that we need a torch. The black emissions from the modem travel through walls to contaminate every room in our home.

Coinciding with these electromagnetic fields we see black contamination trails from our neighbour's wireless. Their wireless signal boosting repeaters and modem emit through our walls and windows. In an apartment the black smoke comes through our floor, roof and walls from neighbouring apartment devices. The closer to the source the higher intensity. Outside we observe black clouds and darkness surrounding a telco tower half a kilometre away. There is no carpet damage. There are no smoked out socks and Saturday night dresses. Yet our home is 'smoked out' by electrosmog.

LEGAL PRECEDENTS

In 1993 *Joseph Criscoula v. Power Authority of the State of New York, 602 N.Y.S 2d 588 (1993)* was based on a 345 kV transmission line (a high voltage line) installation and a potential reduction in property price. The plaintiff settled for damages and legal fees.[31] A couple in Vermont in the US was recently awarded $1 million after a utility, VELCO, located a communications tower on the mountain where they lived and they had to move because of the radiation.[32]

US lawsuits have been lodged by plaintiffs with brain cancer attributed to cell phones. Some were first filed in 2001 and 2002 and many of the plaintiffs are no longer alive. Each plaintiff is asking for over US$100 million. Defendants include companies such as Motorola, Vodafone and agencies and government such as the Institute of Electrical and Electronics Engineers and the Federal Communications Commission. Recently the Washington DC District of Columbia Superior Court ruled five scientific expert witnesses may testify for the plaintiff.[33]

In the Australian case McDonald and Comcare, radiation exposure was recognised in CSIRO employee McDonald's office role. The government health and safety provider Comcare was liable to pay 'compensation in accordance with the *Safety, Rehabilitation and Compensation Act 1988* in respect of an injury, being an aggravation of a condition of nausea, disorientation and headaches'.[34]

People sickened by radiation pollution sickness are often unable to continue working; financial loss often accompanies the decline in health. Taking on the corporate heavyweight legal team of Vodaphone or Apple may not be a consideration when physical and mental health has been decimated. Your focus may rather be making it to tomorrow. My radiation pollution symptoms were so acute during the Sydney EMF Experiment that I took radiation refuge in order to 'make it to tomorrow'. If you choose to pursue legal outcomes to rebalance societal health and harmony know that your actions will be of service to many.

TELCO TOWERS AND PROPERTY PRICES

The internet has enabled a transformation of the working norm. Office workers might opt to work a day a week from their hardwired homes and certain jobs can function entirely from a home office. We could select to live in a low microwave radiation suburb or town. I spent a year immersed in research and writing in exactly this scenario. During the latter part of the 'year at home' I spent one day a week in town for consultation work and other days hardwired and radiation free. As my health improved I correspondingly increased my time in towns and cities talking to groups. One of the questions I was asked by a lady at an event was 'I live next to a tower. Should I sell my house?'

There are variables that will determine the answer but I knew that this lady was moderately radiation pollution sick. My answer was yes. A tower in the neighbourhood will not cause an aggressive decline in property prices but it is just around the corner. A recent meeting on a proposed tower had a progressive demographic with the majority of attendees being young families. A real estate agent with a home in the area warned property prices would fall out of the sky if the tower was installed. Presently a minority of buyers are aware of radiation technologies so they don't ask the question 'Where is the nearest tower?'

When I bought my Sydney rooftop apartment I had not yet been acutely sick and it did not enter my mind to look skyward in four directions. Awareness is starting to shift and I've done measurements for prospective home purchasers interested in minimising their family's radiation exposure. In the near future we may see 10 to 20% sliced off property prices due to proximity to an electromagnetic radiation pollution source. Most people would pay more to live away from a nuclear reactor.

In a recent survey in the US 94% indicated that cell towers and antennas in a neighbourhood or on a building would impact interest in a property and the price they'd pay for it; 79% said they would never purchase or rent a property within a few blocks of a cell tower or antennas.[35]

There is a flipside to the property price dilemma. Creative councils and constituents may choose to declare their town or city centre antenna and Wi-Fi free. In the coming decades these towns might just become Zero EMF boom towns as people seek healthier, simplified lives.

CHAPTER 14

My Journey IV – Frying High

- *I fly to Bali as a last resort for a cure to radiation pollution sickness – to no avail*
- *Learning that present-day electric overload is like being struck by lightning*
- *The worst flight of my life*

IN THESE DAYS OF DISCOUNT AIR TRAVEL I've had my share of memorable rides. There was the time in China when I flew from Xi'an to Beijing after ambling around the Terracotta Army. The flight coincided with China's debut in the football World Cup. This had captured the nation and it seemed 1.2 billion Chinese were glued to their TVs. I was the only passenger on the plane. Service was exquisite. I sat up the front and chatted with flight attendants then stretched my legs across multiple seats to doze. 'This is flying,' I thought as I lay my head on a cushion with an inner smile as if I had just summited Everest. The very next flight it was back to knees around my ears and passengers with body odour. They say descending a peak is the hardest part.

On a plane in America I had a mysterious stench claw at my skin. It came from every direction. I initially blamed it on the heavyweight who'd inhaled his first food tray of turkey sandwiches and demanded a second. Artistry is required to determine a silent farter in a confined space. I writhed and pinched my nose and breathed through my pullover. The passenger next to me remained unperturbed. How could he be so cool? Was he the felon? I generated a double grudge – one on the turkey guy to my left and one for the iceman on my right.

That flight arrived for a stopover and breath of fresh air. Out of Dallas I'd boarded early and with an empty adjacent seat thought, 'This is more like it.' The first thing I saw were the ninja star spurs on his leather boots. I was surprised he got through security. I looked up and saw the leathers. He was straight from a John Wayne movie – proud, loud and friendly. As his volume augmented I became quieter and quieter. I closed my eyes, put my head back and played dead. Halfway to New York I opened my eyes with a start. The cowboy had engaged the passengers behind us. As he swung around his hat sliced me above the eye drawing a trickle of blood. The 'get me off this plane' moments started to shadow me.

After 15 years of frequent international travel I was tired of it but still loved the thrill of a new destination. There was freedom in being able to jump on a plane and go. The Sydney EMF Experiment rewrote my perspective on flying. Experiencing radiation pollution sickness led to a dread of flying. A plane has turbines and extensive electronics and is close to being a Faraday cage. Every signal reflects internally off the metallic shell. Passengers do not always switch off their phones. They tuck their mobile in the back of the seat where it emits mere centimetres from another passenger's spine. On landing within seconds every passenger switches on their device generating an angry spike of electromagnetic radiation and general desperation to escape the aircraft.

Losing the freedom to fly was an inspiration to journey deeper into my health recovery. I'd spent thousands of dollars on self-experimentation, research, products and healing therapies. I had become adept in all things electromagnetic radiation yet continued to experience symptoms when I was around it. I knew there was a final piece for this book and I was guided to research orgonite gifting and energy healing in Bali. After months of avoiding air travel, towers, mobile phones and Wi-Fi I felt ready to fly again and booked my tickets.

At the gate I immediately felt tense and once I was on the aircraft and everyone was sending last minute text messages I felt even tenser. As the plane took off I realised I wasn't ready. To add to the microwave mix the airline had just introduced onboard wireless. All the old symptoms visited. I hit my head against the chair in front of me and this escalated into an argument with the chap seated in it. Thinking I might as well make the most of the ride I pulled out my meter and noted microwave levels of extreme concern per International Building Biology Guidelines (see Appendix E).

The flashing red lights and disconcerting noise emanating from the meter frightened the burly rugby player next to me. The moment I detected his distress I turned to attempt to explain what I was doing but he'd already changed seat. I'd wanted the arm rest but I didn't plan to scare him out of his seat to get it. His biceps were larger than my thighs and with arms that big there was no way to share. A flight attendant came by to ask 'Is everything okay?' I smiled, nodded and stretched myself out over the two seats. Noted for the future was the meter's success as a legroom creator.

After the first leg I arrived in Perth as radiation pollution sick as I'd been during the jittery, headachy days and nights of the Sydney EMF Experiment. Do I continue on to Bali or make this the last flight of my life? I was inspired to find health no matter what the consequences. The journey continued.

HITTING ROCK BOTTOM IN BALI

Bali had been put on a pedestal. The last 12 months had smashed my life to pieces and Bali was a last hope and in my mind a place where miracles occur. At Perth Airport I was selected from hundreds to go through the body scanner. The model they used was based on electromagnetic radiation. I tested my rights and refused to go through. A commotion ensued and the security manager arrived. I informed him that I was electromagnetic hypersensitive and would not traverse the body scan. He told me the complaint was a new one for him which was interesting to me but on reflection made sense; people experiencing radiation pollution sickness wisely avoided international flights. I signed a document and went through the standard aisle.

The flight to Bali was not as intense as the onboard wireless flight cross-country. Soon my head was boiling, though, as I stepped into a two-hour customs queue of one thousand mobile toting tourists. Bali was a radiation refugee nightmare on par with inner city Sydney. I continued testing products promising to counter toxic energies to no avail. The low altitude of mobile phone towers in Bali meant direct trajectory was not much higher than head height on a motorbike. This increased the health impact of towers. I had four towers within a kilometre of my room in a relatively remote part of south Bali (Bukit). After a sleepless night I moved.

I hadn't found the elusive cure for radiation pollution sickness. In Bali I engaged reputed energy healers to help crack the code. After a few sessions of deep tissue massage and energy work I felt slightly better but it only took a motorbike ride back to my accommodation to again feel kryptonite-drained. I was so desperate that I sat in the chair for intravenous ozone therapy. I opted out last minute after the therapist revealed a dilapidated drip hanging from a corroded stand more suitable as a prop in a Mad Max movie. My next stop was one of the island's famed Ayurvedic healers.

AYURVEDA AND THE POISON OF INDRA'S THUNDERBOLT

In the north of Australia the wet season brings thunderstorms with week-long electromagnetic preludes. The tension builds as thunder wants to roar and rain wants to pour but both hesitate. I've found myself more irritable and anxious during these periods. Symptoms were similar to but not as severe as radiation pollution sickness.

I had a basic understanding of Ayurveda and guided the Bali practitioner to check my pulse. I indicated how it became jittery with radiation exposure and would fade to almost undetectable. His disinterest reminded me of the GP in Sydney. After taking off my shirt for another oily back massage I knew I was no closer to a miracle.

Perhaps I needed someone from the Raj Vaidya lineage? R.K. Mishra from the Raj Vaidya lineage of physicians talked about seeing his patients in London. He said most had a disturbed laya (rhythm) in their pulse. He wasn't sure of the cause so called his father, who was a more experienced practitioner of the same lineage. His father said he knew of the described laya and the lineage had been treating for it prior to modernity –

'The laya I felt in the London pulse readings was similar to problems experienced by people when they or their residence had been struck by lightning. Present-day electric (over-) load can create problems akin to lightning in the pulse.'

His father's instructions were to 'follow the same protocol … used on people who had been struck by lightning … After contemplating his answer, I realised that this disturbance involved vibrational pollution or toxicity corrupting the physiology through the channels and disturbing the marmas. Marmas are vital points, trigger points for pranic flow… The primary area affected is the sushumna nadi, through which these corrupt energies enter the spinal cord and the physical body… The primordial force or energy of nature in the form of prana also continuously enters through the nadis (vibrational channels). Exposure to EMF can pollute the nadis, the vibrational channels which in turn, can disturb the physical channels and create imbalances in the physiology's entire cellular network.'[1]

From an Ayurvedic perspective the symptoms of those in monsoonal India (where they call it Poison of Indra's Thunderbolt) include insomnia, anxiety and depression and have the same root as radiation pollution sickness.

My Bali healer experience had failed. I now wanted to experiment with orgonite. Could a chunk of iron filings in resin with a few crystals embedded change my life and allow me to shake off my radiation refugee status?

FLYING HOME AND SURRENDERING

Don Croft and others took Reich's theories around orgone and developed Tower Busters, Holy Handgrenades, ChemBusters and other orgonite products to counter electromagnetic radiation and other forms of pollution. In Bali I met a gifter who made his own orgonite pieces and generously gifted me six Tower Busters. Might orgonite be the solution I'd spent so long searching for? I felt a surge of hope. The theory is one or two TBs will transform the 'deadly' energies of a microwave tower.

A Tower Buster is about the size of a teacup. I decided to put one around my neck. I had it drilled out and some leather attached and I began to wear it while riding my motorbike around the island. Another one was kept in my backpack and the other four I used at each corner of my room. They had a noticeable effect and certainly aided my motorbike excursions in that I did not feel as drained of energy. However, something remained off with my subtle energetics. I felt less radiation affected but continued to feel ill and irritated.

As days passed the likelihood of an anticipated healing diminished. The hopium wore off and depression resumed. Under every stone I lifted on the journey remained the inescapable muddiness of radiation pollution sickness. I now just wanted to get away from the toxic energies that were as intense in Bali as in inner city Sydney.

The Bali trip was a turning point in my journey. I'd exhausted my search for a magic pill solution. I'd tested every product. Even my interactions with the famed Bali healers had been ineffective. Flight 206 was the very worst flight of my life and to this day the last I have taken. It was clear I had to stop flying. It was time to stop searching for an elixir. The moment of surrender prior to landing brought profound clarity and relief. I now had only one place left to go. In the second phase of my journey that place was inside myself.

CHAPTER 15

Electromagnetic Humans

- *Energy fields and chakras in those experiencing radiation pollution sickness*
- *The heart as our electromagnetic field generator*
- *How our children may evolve in a microwaved society*

PSYCHIC AND RADIATION POLLUTION SUSCEPTIBILITY

IN CHAPTER 2 I DISCUSSED THE FOUR E'S that lead to radiation pollution sickness. Empathic people may pick up and 'take on' other energies. If a friend was experiencing grief over a relationship I could meet with them and unconsciously take on their burden. After half an hour they would be chirping like a bird and I'd be left carrying the weight of their grief. This is energetic contamination or psychic pollution. I either had to be more selective in the people I met or work to establish a stronger and less penetrable energy field. If we are susceptible to psychic contamination we are more susceptible to energetic contamination from electromagnetic radiation. I'd known for years about my empathy and tendency to take on other people's pain and burdens but I hadn't worked out how to work with it. Sometimes we only take action when the pain and loss is significant enough to create an impetus. We only move from a house we've been unhappy in for years when a tree falls on top of it. After the losses during the Sydney EMF Experiment I decided to do what it took to strengthen my energy field. I started with what I knew from yogic traditions and chakra anatomy.

WORKING ON MY CHAKRAS

Chakras are functional energy vortices each with an anatomical reference (see Appendix G and Figure 16). They allow energy transfer via the subtle energy field and can be:

- Depleted or weak
- Receiving
- Activated
- Overactivated

The sweet spot is 'activated' so they transmit and receive like healthy antennas. Receiving chakras take in energetic information but don't transmit, so like wallflowers we observe but nobody knows we are there. A depleted chakra means that area of our anatomy and the associated energetic aspect are shut down. For instance, a depleted second sacral chakra may be indicated by shut down creativity and sexuality, and lower back problems. After a while the depleted chakra will 'lash out' and compensate with a moment of overactivation. In the example this might be addictive compulsive behaviour. The metaphysical phenomena of chakras has been scientifically verified. In one test visible light was generated from the heart chakra.[2]

Key chakras for those of us experiencing radiation pollution sickness are the third solar plexus chakra and the fourth heart chakra. With a depleted third chakra we relate to the role of victim. We attract someone with an overactivated third chakra (perpetrator or bully) to compensate. It might be a work or intimate relationship where we feel used and someone indifferently uses us.

Not long after moving into the apartment in Sydney I had the building Owners Corporation and lawyers pursuing a strata non-compliance. I had inherited the issue but was unaware of it when I bought the apartment. The previous owner had bolted overseas. In proceedings it came out that the Owners Corporation had known about the non-compliance for years. Why had they not taken action earlier? A 90-year-old pensioner is more likely to have her bag snatched than a 25-year-old rugby player. They'd not taken action with the 'rugby player' previous owner, but when I came onto the scene the Owners Corporation sensed vulnerability. I was 'a pensioner with a handbag' for their overactivated third chakra or perpetrator energy attacks.

The affair was perfect for 60 MINUTES and in a society where we elevate suffering and the role of victim I briefly contemplated media involvement. But I saw how I played my dysfunctional part and the losses provided impetus for internal transformation. It was time to strengthen my energy field so I could interact in the world without being drained by vampires.

I was consulting with clients only one day a week because I felt so sapped of energy when I spent time around technology and people. Jane was experiencing radiation pollution sickness and she lived life as a victim. She was in an abusive relationship. She always ended up losing money on business interactions and her work in a healing profession left her exhausted. Jane almost walked out of my office when I told her she had to get over being a victim or nothing would change. I clarified that abuse from perpetrators was unacceptable but it was up to her to make that clear by transforming her energy field and cutting ties to those relationships.

Jane had been familiar with disempowering sympathetic responses like 'oh poor dear' rather than empowering truth. She was at first a little ruffled by what I said. When I shared the parallels of my own journey she got it. Jane was living a life in hiding to avoid psychic pollution and electromagnetic pollution as I was. I decided that both of us needed to start chakra activating yoga.

Jane and I required third chakra solar plexus (abdominal) strengthening. The supermarket meat section fluoro lights in gyms had put me off. Sweaty six pack abs were for others. The great thing with a yoga solar plexus activation is you only need one minute and you can do it anywhere:

Lying on the floor or mat in stretch pose lift your feet 20 cm off the floor with toes pointed. Head, shoulders and upper body come off the floor and arms are straight with hands facing the sides of your legs and fingertips stretching towards your toes. The lower back remains on the floor or you can place your hands palms down at the lower back for more support. Take your awareness to the abdominals and breathe vigorously through your nose as you pump the area. The dristi or eye gaze is to the big toes. A minute per day is enough. The exercise is not to be performed on the heaviest days of menstruation.

This simple yoga pose works wonders as part of a holistic approach to reduce susceptibility to psychic and electromagnetic pollution.

Now on to the vital fourth heart chakra. Why are so many songs and poems written with language of the heart? 'He broke my heart.' 'My heart skipped a beat.' 'A teardrop caressed my heart.' A sportsperson can lose a game but it doesn't matter because 'she played with heart'. The heart is an intelligent electromagnetic field powerhouse.

POWER OF THE HEART

The heart produces an electromagnetic field 5000 times as powerful as that produced by the brain.[3] This field projects from your physical body to contribute to the circumvent field or aura. Sixty to sixty-five per cent of heart cells are neural cells connected to the central nervous system and brain and concomitant with the amygdala, thalamus, hippocampus and cortex.[4] Heart transplant recipients show a different personality post-operation. An ill-tempered boss has a month off work for a transplant and comes back a different person. A change in heart may lead to a 'change in heart'.

The heart is a dynamic, non-linear harmonic oscillator.[5] The heart's pacemaker cells are biological oscillators that generate an electric field. Simultaneously the heart produces a magnetic field which forms a pattern similar to a standard magnet.[6] This electromagnetic field is constantly shifting. The heart produces a spectrum of electromagnetic frequencies with vast amounts of information contained in each frequency. As with a holograph each segment of the heart's field holds all information for the whole field.[7] This field is torus shaped per all living organisms including the cells of our body, plants and the earth.

As a non-linear oscillator, the heart is extremely sensitive to disturbance of its dynamic equilibrium. Heart cells entrain (or synchronise) to beat in unison with adjacent cells.[8] Entrainment occurs in-utero and during infancy. The field of an infant is entrained through the mother's electromagnetic field. This is the basis for the developmental psychology understanding of the impact of the mother's emotional state during pregnancy.

Institute of HeartMath scientists here describe direct knowing and intuition – 'The intelligence of the heart ... processes information in a ... more intuitive and direct way ... shows us the inherent core values in our lives and brings us closer to the sense of true security and belonging ... often accompanied by a solid, secure, and balanced feeling ... by a peaceful, clear state of awareness.'[9] These qualities are antonym to radiation pollution sickness symptoms. One of the key 'losses' during the Sydney EMF Experiment was my ability to 'know' the answer and connect with intuition. My decision-making processes were therefore distorted.

HeartMath has found the heart communicates with the brain and rest of the body –

- Neurologically – Through transmission of nerve impulses.
- Biochemically – Through hormones and neurotransmitters.
- Biophysically – Through pressure waves.
- Energetically – Through electromagnetic field interactions.[10]

'Just as cell phones and radio stations transmit information via an electromagnetic field, recent research has led some scientists to propose that a similar information transfer process occurs via the electromagnetic field produced by the heart.' Additionally, the heart's electromagnetic field can be measured up to 10 feet away.[11]

Our heart beats incoherently when feeling frustration versus a coherent waveform (or beat) when we feel appreciation (see Figure 9). Another word for this incoherence or chaos is noise. If someone is emanating frustration we 'pick up' their electromagnetic field. Similarly if they had a melodious day and are feeling appreciative we pick up their coherent waveforms. We might ask them what wonderful thing happened at work to make them so buoyant.

My experience suggests our electromagnetic field can be detected further than the Institute of HeartMath evidenced 10 feet. Have you been at a party on the far side of a large room when a radiant woman or charismatic man enters? Despite having your back to them you feel their presence 20 metres away (over 60 feet) and are compelled to turn your head towards them.

The chaotic heart waveforms of frustration in Figure 9 are representative of a response to internally or externally generated psychic pollution. When we pick up the incoherent electromagnetic field of our partner our emotional state is influenced. If our electromagnetic field is optimally healthy and in appreciation – as per Figure 9 – then our electromagnetic field harmonises their noise. If on the other hand we are not functioning optimally then their incoherent noise field interferes with our electromagnetic field. Our field is overpowered by their field of frustration and we become frustrated. An incoherent field is a field of arguments.

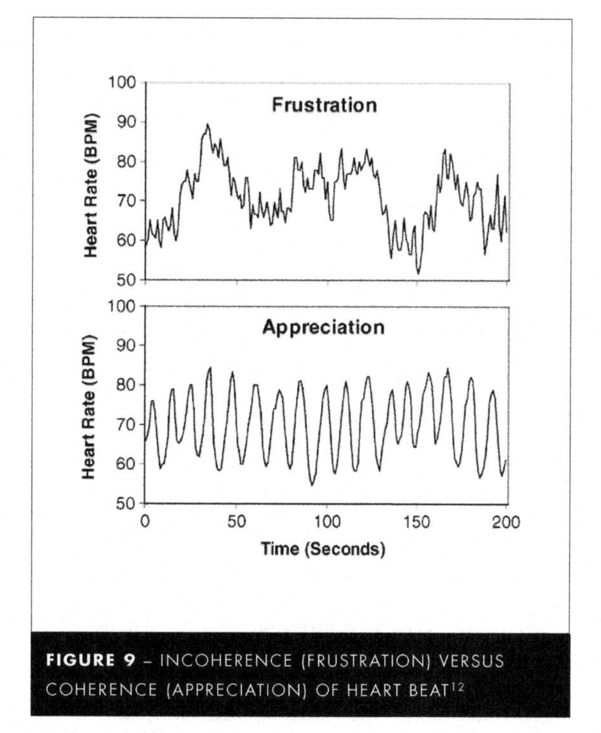

FIGURE 9 – INCOHERENCE (FRUSTRATION) VERSUS COHERENCE (APPRECIATION) OF HEART BEAT[12]

Microwave towers are a frustrated partner equivalent. They generate incoherent (noise) shaped waveforms at multiple frequencies. We receive and process these electromagnetic fields in the same way we receive the electromagnetic field of our partner. The Sydney EMF Experiment corresponded with living with a frustrated or chaos waveform generating partner 24/7. I was not in an optimally healthy appreciative state to generate coherent waveforms with sufficient energy to harmonise the incoherent electromagnetic radiation I had to process. The result was I was overwhelmed by the noise fields projected by towers and devices and constantly felt myself frustrated and irritated at body-mind and cellular levels. This is a form of mind control as expanded on in Part VI.

At rest individuals have their own unique signature fields which are a full spectrum of smooth, coherent waveforms. The electromagnetic waveforms that Peter generates are different from those generated by Jennifer. A study confirmed disease or disease that was about to develop showed on the signature field as jagged incoherent waveforms concentrated in the low and high frequency bands. 'Deficiency pattern' diseases such as cancer and chronic fatigue syndrome showed up as incoherent patterns in the high frequencies. 'Hyperactive pattern' conditions such as hypertension and skin problems showed as incoherence at low frequencies.[13]

Are the incoherent waveforms associated with disease showing in our energy fields due to radiation emitting technology? If our own energy field is influenced by discordance this may inform future illness. The opportunity is to heal our energy fields in order to heal on the physical plane.

THE EVOLUTION OF ELECTROMAGNETIC CHILDREN

Children born today have not known anything other than radiation levels millions of times higher than those Generation X and older were brought up in. From the moment of conception they have developed alongside mobile towers, mobile phones and wireless. The precautionary principle has been abandoned and we have no idea where this worldwide electromagnetic radiation experiment is leading the health of our children. Scientists around the world are speaking out to recommend we urgently re-evaluate exposure limits.[14]

Radiation exposure causes adverse biological effects to adults but are our children more or less affected? It may be that adverse health effects for children are more significant than for adults. Children have smaller skulls and bodies and their organs are more readily penetrated by microwaves as we see in a later chapter. It may also be that children are evolving with such fluidity that their systems have mechanisms to handle electromagnetic radiation that adults don't have. An evolved system of intelligence may endow them with a capacity to harmonise the same electromagnetic radiation that debilitates adults.

CHAPTER 16

Body–Mind as an Antenna

- *How our glands, organs, brain and cells act as potential antennas*
- *Why constant background noise stops you from tuning in to yourself*
- *The fractal nature of our DNA means it acts as a receiving antenna of all wavelengths of radiation*

MUSICAL INSTRUMENTS COME IN VARIOUS SHAPES AND sizes to resonate a certain frequency of sound waves. A violin is a different shape and size to a cello. A trumpet is a different size and shape to a piccolo flute. Similarly antennas come in different shapes and sizes to resonate and amplify a certain radiofrequency. Antenna lengths of one wavelength λ, $\frac{1}{2}\lambda$ and $\frac{1}{4}\lambda$ are important in basic antenna theory. A common ½ wave dipole antenna is the 'rabbit ears' antenna we shuffled on the top of our TV for a better signal during the 80s. Another example is the common centre-fed symmetric T antenna as found on top of Australian roofs in which each 'ear' of the antenna is $\frac{1}{4}\lambda$, combining to make $\frac{1}{2}\lambda$.

Our glands, organs, eyeballs, brain, cells and DNA strands are potential antennas. They are different sizes to resonate with a certain wavelengths. Table 1 displays a handful of common telco frequencies used in Australia and their associated wavelengths. The ear to ear measurement of my own head is approximately 16.5 cm which may amplify 900 MHz and 1800 MHz microwaves. My eyeball measured just under 3 cm across – perfect for amplification of 2600 MHz and 5 GHz microwaves. My hypothalamus would be around 1.5 cm across – an ideal antenna for 5

GHz Wi-Fi. We could take this analysis a step further to correlate symptoms with the frequency band of exposure. A certain frequency may have a larger impact on melatonin levels for example, due to resonance with our pineal gland antenna.

TABLE 1 – FREQUENCIES AND ASSOCIATED WAVELENGTHS FOR ANTENNA DESIGN - λ, ½ λ, ¼ λ

FREQUENCY BAND	λ	½ λ	¼ λ
700 MHZ	42.9 CM	21.4 CM	10.7 CM
900 MHZ	33.3 CM	16.6 CM	8.3 CM
1800 MHZ	16.6 CM	8.3 CM	4.2 CM
2600 MHZ	11.5 CM	5.8 CM	2.9 CM
5 GHZ	6.0 CM	3.0 CM	1.5 CM

TUNING IN A BACKGROUND OF NOISE

We would slide the antenna around the top of the TV and adjust the length of the antenna arms to tune it. Background noise would emanate from the TV but then we would find the sweet spot where it became a clear, crisp signal. Considering psychic pollution, electromagnetic radiation pollution and chemical pollution as background noise our own authentic signal is the TV channel we want to tune to. During the Sydney EMF Experiment I 'did not feel my Self' and struggled to tune to my own channel for two reasons –

1 My own transmitted electromagnetic field signal was weak.

2 There were strong electromagnetic radiation pollution and psychic pollution signals overlaying..

Part of the reason for the yoga boom is that people want to get back in touch (tune in) with themselves. Yoga, dance, meditation or sports can strengthen our electromagnetic field and simultaneously quieten background psychic and electromagnetic noise. However, we can still feel out of touch despite these activities. During the Sydney EMF Experiment my symptoms were so severe that even with yoga and meditation I was not able to tune in to my own clear, crisp signal.

My electromagnetic signature field signal (related to my unique DNA blueprint)

was noise affected. The miraculous fractal functionality of our DNA also makes it susceptible to electromagnetic radiation pollution.

FRACTAL DNA

A fractal is a pattern that repeats with self-similarity so that on multiple scales the shape is the same. Lightning bolts, rivers, fjords, shorelines, trees, mountain ranges, clouds, broccoli and sea shells are imbued with fractal patterning.

DNA is a molecule that keeps us functioning and makes the proteins we need to live. At 2 m long and with 3 billion base pairs the double helix coils up in a 'coiled coil' in the nucleus of a cell. Like a coiled ladder of base pair steps it functions as a fractal antenna therefore responding to a wide spectrum of frequencies. This is the Holy Grail for antenna engineers. A centre-fed symmetric T antenna would need millions of metallic T-lengths to even compare with characteristics of a single DNA double helix coil. The fractal nature of DNA means it acts as a receiving antenna for all common microwave frequencies including those in Table 1. This functional perfection also puts DNA at higher risk of radiation damage because it has the capacity to be an antenna and therefore process all wavelengths of emitted radiation.

Scientist and researcher Henry Lai refers to a Blank and Goodman paper for an explanation of similar genetic effects for a range of frequencies. The 'wide frequency range of interaction with EMF is the functional characteristic of a fractal antenna, and DNA appears to possess the two structural characteristics of fractal antennas, electronic conduction and self-symmetry'.[15] Due to fractal antenna characteristics of DNA we are susceptible to genotoxic effects from any wavelength in the spectrum whether nuclear bombs, X-rays, the microwaves of Wi-Fi and telco communications or extremely low frequency electricity and appliances.

CHAPTER 17

Earth Energies

- *Examining how the Earth's energies can contribute to overstimulation, especially in our sleeping areas*
- *In Europe, earth energies have long been recognised as a source of illness*
- *Why smart phone blue screens disrupt sleep*

WHILE DETOXING IN THE ZERO EMF SANCTUARY the quality of my sleep was moderate to poor. Some nights I'd toss and turn until morning and something felt energetically awry. Weeks later I camped at an eco-village community and was astonished to wake as refreshed as I could ever remember. The miracle sleep occurred for five straight nights. Why didn't I sleep like this at home? The camp spot was in a near zero mobile reception zone and I'd not used my mobile phone or computer during those days. That was the first time in many years that I'd switched off for an extended period. The Zero EMF Sanctuary had similarly limited mobile reception yet I didn't sleep nearly as well. Either the land at the eco-village was conducive to sleep or my computer use (hardwired mouse and keyboard) and rare mobile phone use at home was enough to cause sleep disruption. The energy had been so calm in my tiny dome tent.

I'd divined (dowsed) in the past to find water. Could the land at the eco-village have a low stimulation and therefore properties for deep sleep? Hoffman-La Roche (Roche pharmaceuticals) in Switzerland employs dowsers to find water for their chemical processes saying, 'Roche uses methods that are profitable, whether they are scientific or not.'[16] Using a similar technique, on my next trip to the eco-village I measured geopathic stresses of the earth. The one-and-a-half man tent had been just outside stimulating earth energies of water veins, Hartmann lines, Curry lines, ground faults and ground mixing. The tent's compactness had prevented me from rolling into these zones. At home the geopathic stress lines intersected right where my head lay on the pillow. In my half sleep I would roll all over my bed seeking a suitable spot with 'quieter' energies. Due to the criss-crossing of these earth energies there were no spots on my bed where I could stretch out and not be overstimulated.

We've seen earth energies in action with mushrooms and medicinal herbs thriving above water veins. Horses, cattle, dogs and chickens avoid geopathic stress. Cats seek geopathic stress and have a go-to spot with higher earth radiation for their nap. Bees seek earth radiation for their hives. Humans only require a small amount of these energies. A constantly crying baby may have her cot located over the intersection of these geopathic stress lines but moving it two or three feet may transform her (and your!) sleep habits.

EARTH ENERGIES IN EUROPE

Countries including Russia, Poland, Germany and Austria recognise the link of earth energies to health. German doctors ask patients where they sleep and for how long they have slept there. They correlate this with the timing of illness. They may appoint a geobiologist to review earth energies at a patient's residence. Different measurement scales exist. German doctor Gustav Freiherr von Pohl introduced the scale 1–16 with levels above 9 being cancer inducing.[17]

To reduce geopathic stress we can change our sleep position. We could move house or become a gypsy. A 1990 study on the health of gypsy travellers in Europe found their cancer incidence to be the lowest in the Western world at 0.6%. This was despite drinking, smoking and exercise habits being no better than the rest of the West.[18] In Australia in 2009 the risk of developing cancer before 85 years old was 1 in 2 for men and 1 in 3 for women.[19]

In 1929 von Pohl produced a map of all the underground waterways in his village. Another map showed locations of cancer deaths. The maps matched with all 54 deaths occurring in locations above water veins. Sceptics asked for a repeat study. In another location the same result was obtained.[20] Käthe Bachler found so many correlations with earth energy and behavioural issues in children that the Education Department in Salzberg gave her a research grant and adopted her 'rolling classroom' system whereby students would change seat every few weeks.[21]

Not everyone wants to lead a gypsy lifestyle. Physical solutions reduce earth energies using harmonic or disharmonic oscillation circuits. I used physical solutions for a period but discovered they aren't necessary and developed my own methods for clients to harmonise their sleeping and living areas.

THE SMART PHONE BLUES

Earth energies can disrupt sleep quality by overstimulating us. This is common at the intersection of gridlines and water veins for instance. One of the ways electromagnetic radiation disrupts sleep quality is through disturbing melatonin production. Smart phone blue screens do the same. A study showed saliva melatonin levels significantly reduced after blue light (short wavelength) exposure.[22]

The Mayo Clinic suggests reducing light levels to below 30 lux which is the threshold thought to suppress melatonin production. An iPhone 4 at 14 inches gave off 8 lux at maximum brightness and < 1 lux at minimum brightness. At 0 inches it gave off 275 lux at maximum brightness and 2 lux at minimum brightness.[23] Holding the phone at arm's length and switching it off an hour before bed may lead to dramatic improvements in sleep quality.

CHAPTER 18

The Medical Community's Treatment of EMR

- *The combination of EMR and drugs is a new frontier*
- *How mobile phone use is implicated in brain cancers*
- *Electromagnetic radiation's effects on the pineal and pituitary glands*
- *Despite the evidence the mobile industry says there is no link between mobile phone use and brain cancer*
- *Royal Raymond Rife's study of radiofrequency waves to 'devitalise' tuberculosis and cure cancer*
- *Georges Lakhovsky's controversial medical treatment invention, the Multiple Wave Oscillator, and his concern of the danger of microwaves*

DRUG AND EMR INTERACTIONS

WHEN IT COMES TO COMBINING FORMS OF energy (or drugs being an energy form), linear science is overwhelmed. Non-linear reactions occur such that the effects of one drug may be annulled or amplified by another. Side effects may be completely different to the common side effects when taking one drug alone. Doctors admit to not fully understanding drug–drug interactions. What happens when we combine alcohol and antibiotics with antidepressants? Regularly we hear of celebrity deaths attributed to toxic combinations of approved medications. Most Australians swallow or inject something with 91% of those 65 years or over taking medication.[24]

Drugs are a chemical energy that have biological effects on humans. A group of physicians developed a system to rate benefits versus harm for various drugs. For example, for participants taking statin drugs over a five-year period results showed –

- None had their life saved
- 1 in 104 were helped (prevent non-fatal heart attack)
- 1 in 154 were helped (prevent stroke)
- 1 in 50 were harmed (developed diabetes)
- 1 in 10 were harmed (muscle damage)[25]

Every drug has a benefit and a consequence. Anti-inflammatories temporarily reduce inflammation. They may induce dizziness, diarrhoea and headaches yet benefits remain alluring.

Electromagnetic radiation is an energy that has biological effects on humans. Benefits versus harm for various forms of electromagnetic radiation include –

- Facilitate ease of communication
- Dissemination of knowledge available on the World Wide Web
- An emergency device to call 000
- 3–5% experience electromagnetic hypersensitivity (including headaches, insomnia, heart palpitations, brain fog, immune depletion and depression)
- 35% may experience mild to moderate symptoms of electromagnetic hypersensitivity

- Have to leave workplace
- Decline in freedom and quality of life
- Social consequences such as online bullying, sexualisation of society, social withdrawal

Drug–drug energy interactions are little understood, but EMR–EMR energy interactions are an unexplored frontier in medical science. What are the biological effects on our body-mind of telco tower microwaves at 1800 MHz in combination with a 5 GHz Wi-Fi signal and 50 Hz electrical field? Adverse biological effects may exceed those attributed to exposure to a single frequency (say Wi-Fi) alone. Similarly each mobile telephony technology has a form of transmission and modulation that is variable. Almost 10 years ago on the transition from 2G (GSM) to 3G (UMTS) Franz Adlkofer announced his latest research results 'There is no doubt – UMTS is ten times more damaging to genes than GSM radiation.'[26] We now have 4G (LTE) interspersed with 3G antennas and 5G is just around the corner.

In the context of EMR-drug interaction we might call a drug any form of energy that when absorbed has a biological effect on our body-mind system – this includes chemical and environmental toxins. Exposure to mould (an environmental fungus) causes specific biological effects. Electromagnetic radiation exposure causes specific biological effects. During a brief exposure to mould and electromagnetic radiation I observed a magnification of my radiation pollution sickness symptoms. The specific symptoms that amplified might have been different if I'd consumed antidepressants rather than absorbed mould spores. The EMR-drug interaction would again be different if the 'drug' combined with microwaves was fluoride in water. The biological mechanisms of combined energies are too advanced for current science. Yet, like lab animals, we continue our part in the microwave experiment.

THE ROCK STARS OF NEUROSCIENCE

If neurons are the lead singer or guitarist of neuroscience then glial cells (neuroglia or glia) are the drummer thumping away rhythmically in the background. Until recently they were considered mere scaffolding. Research indicates glial cells are support cells for the central nervous system and signal the neurons that influence our breathing.[27] Astroglia are a type of glial cell also referred to as astrocytes. JJ

Rodriguez and fellow researchers have linked astroglia to Alzheimer's disease and dementia. 'Astrocytes, the most numerous cells in the brain, weave the canvas of the grey matter and act as the main element of the homoeostatic system of the brain. They shape the microarchitecture of the brain, form neuronal-glial-vascular units, regulate the blood–brain barrier, control microenvironment of the central nervous system and defend the nervous system against multitude of insults.'[28]

Mobile phone use is implicated in brain cancer in the form of glioma (glial cell tumours).[29] Edgar and Sibille indicate altered glial structure and function is involved in several mental illnesses. Reduced oligodendrocyte (type of glial cell) function is linked to these illnesses with similar features in models of stress induced depressive-like biophysical behaviours such as chronic stress.[30]

The above findings substantiate my own mental health deterioration and depression during the Sydney EMF Experiment. Electromagnetic radiation may generate similar adverse health effects to chronic stress and we need to address both in order to re-establish physical and mental health.

Post-mortem analysis of the brain tissue of depressed individuals indicates glial cells are reduced.[31] Conceivably mental illness and the neurological decline of dementia and Alzheimer's disease are linked to electromagnetic radiation exposure –

EMR exposure > altered neurological structure > mental illness, dementia, Alzheimer's disease

THE MASTER GLANDS

The pineal gland is tucked deep inside the skull and has blood flow second only to that of the kidney.[32] It produces the serotonin derived hormone melatonin, crucial for sleep modulation and creates biochemicals tryptophan, tryptamine and pinoline. Due to structural association to melatonin DMT (dimethyltryptamine) may be made in the pineal. DMT has been described as the Spirit Molecule and is connected to mystical experiences and visions reported by near death experience survivors. Recently DMT (dimethyltryptamine) was found in the pineal gland of live rats[33] to supplement this association.

The psychic aspect of the sixth or third eye chakra is linked by yogis to the physical pineal gland. The pineal is also linked to the seventh crown chakra, which is known as a gateway to spiritual connection and wisdom. The pineal gland production of melatonin is reduced by exposure to extremely low frequency magnetic fields.[34] Electromagnetic radiation's effects on the pineal gland therefore have profound implications not only for our sleep and mood quality but our spiritual connection.

Medical doctors seek to bring the body back to homeostasis or stability. The pituitary gland is our body's inbuilt medical doctor. It secretes hormones that regulate homeostasis including body temperature, pH levels, water levels through the kidneys and blood glucose. The pituitary gland is part of the hypothalamic-pituitary-adrenal axis (HPA) that is involved with stress, immunity, digestion, mood, and energy regulation. Like the pineal gland it is safely tucked deep inside our skull for good reason.

In the Sydney EMF Experiment one of the surprising symptoms that arose alongside adverse biological effects was how I lost touch with spiritual connection. I had a sense of confusion, separation and being without guidance. I stopped attending yoga and meditation as it did nothing to alleviate my disconnection. Yogis are aware of this distorted state and link it functionally to the sixth and seventh chakras which are anatomically connected to the pineal and pituitary glands. This experience of spiritual separation, even when I was participating in a 'spiritual' activity, led me to coin the terms 'spiritual effects' of electromagnetic radiation.

THE LINKS BETWEEN MOBILE PHONE USE AND BRAIN CANCER

The World Cancer Report 2014 says cancer has taken over from heart disease as Australia's biggest killer. Brain cancer is the leading cause of cancer death for those in Australia under 39.[35] In 2010 WHO and the International Agency for Research on Cancer performed a study with a significant proportion of funding coming from the Mobile Manufacturers Forum and GSM Association. The majority of the subjects were light phone users of 2 to 2.5 hours per month. The heavier users talked for 1640 lifetime hours over 10 years or approximately 30 minutes per day. These users had ipsilateral (same side of the head) tumours.[36]

A study by local neurosurgeon Vini Khurana and co-resesarchers links prolonged mobile phone use with ipsilateral brain tumours. 'The results indicate that using a cell phone for > or = 10 years approximately doubles the risk of being diagnosed with a brain tumour on the same (ipsilateral) side of the head as that preferred for cell phone use. The data achieve statistical significance for glioma and acoustic neuroma but not for meningioma.'[37] A year after the Interphone study the WHO 'classified radiofrequency electromagnetic fields as possibly carcinogenic to humans (Group 2B) based on an increased risk of glioma, a malignant type of brain cancer, associated with wireless phone use.'[38]

Australian neurosurgeon Charlie Teo sees brain cancer first hand. Dr Teo said, 'I see 10 to 20 new patients each week and at least one third of those patients' tumours are in the area of the brain around the ear. As a neurosurgeon I cannot ignore this fact…'[39] The body representing Australian mobile industry interests responded – 'To date, no adverse health effects have been established as being caused by mobile phone use.' And 'there is no substantiated evidence of an increase in brain cancer and no association between brain tumour risk and mobile phone use'.[40] This response contradicts the WHO classification of 'possibly carcinogenic' a year earlier. It reminds me of a child who has broken a window playing football. The neighbour saw it all happen but the child is too afraid to admit the truth. 'Did you kick the ball into the window?' 'No.' 'Then how did the window break?' 'Wasn't me.'

The WHO says cancer cases will surge by more than 70%. Their report calls for taxes, health warnings and advertising restrictions on alcohol and junk food to protect the world against the disease.[41] I know healthy 60, 70 and 80 year olds who drink moderately and enjoy regular deep fried fish and chips. Regulatory bodies distract us from what is really going on. The role of the stress response and DNA damage in cancer (per Appendix B) has been ignored until now. Psychic, electromagnetic radiation and chemical pollution are all overlooked stress response inducing root causes of cancer. In combination they contribute to the WHO's predicted 70% surge in cancer unequivocally more than a greasy hamburger with bacon.

MISLED BY THE MEDICAL ESTABLISHMENT – THE ECT SCAM AND DIATHERMY

Medical advice on new technologies can be dangerously misleading. Electroconvulsive therapy (ECT) uses electrically induced seizures to halt depression. There is not a single study to show an improvement in a patient one month post ECT treatment.[42] Patients experience ongoing cognitive effects which include 'deficits' such as slowing of reaction time and retrograde amnesia as a result of the therapy.

These cognitive effects were never studied until 2007 despite the therapy existing for over half a century.[43]

Female physiotherapists who use shortwave and microwave diathermy units increase their odds of miscarriage.[44] Diathermy is the use of electrically or electromagnetically derived heat for muscle therapy or cosmetic and skin procedures. The sky has become one big microwave diathermy unit. Pregnancy and fertility groups might factor the escalation of microwave exposure into their discussion of root causes of miscarriage.

RIFE AND OSCILLATION RATES OF CANCER

Raymond Rife was a scientist inventor who in the 1920s and 30s sought a cure for cancer. He tested five models of an incredibly powerful microscope that used natural light with up to 50,000 power (magnification). A standard light laboratory microscope has 1,500 to 2000 magnification. We now have electron microscopes with equivalent power but they use short wave radiation which is detrimental to the micro-organism being viewed due to X-ray bombardment. Rife's microscope allowed micro-organisms to be viewed in-vivo.

His microscope was unique in that light was bent and polarised such that a sample could be illuminated by a single light frequency. Rife could then resonate the light of the microscope with that of the sample micro-organism to determine a colour and frequency at which the micro-organism was vibrating. In the 1920s Rife went to work viewing tens of thousands of cancer cultures. The problem he had was obtaining the cancer microbe in his test cultures. By experimenting he found microbes respond to the light of noble gases like neon and argon.

Rife set up a test tube with cancer tissue in a closed loop with argon gas. The gas was charged electrically and Rife observed a cloudiness linked to ionisation. He then took the tube and put it in a water vacuum at body temperature for 24 hours. Viewing the tube under his microscope something smaller than bacteria emanated red/purple light. He called these BX (Bacillus X). Rife had connected the beyond visible oscillation rate of bacteria to the visible spectrum.

Rife spent decades experimenting and refining. He found that by changing by two parts per million the medium in which BX was living it would change to a new enlarged form he called BY (Bacillus Y). He linked his discovery that micro-organisms light up at a certain colour when activated by a light frequency to the possibility that those micro-organisms could be 'devitalised' by frequencies also. Hence the frequency generator with the potential, knowing the frequency of the virus, to 'devitalise' tuberculosis and typhoid.

Rife reported to the special medical Research Committee of the University of Southern California. Sixteen hopeless cases of malignant cancer were treated. A pathologist and five doctors signed off fourteen cases after three months. They were cured with just three-minute treatments at three-day intervals with the frequency generator set at the oscillatory rate for BX. Word spread and Rife's world began to implode. Healing technologies that threaten existing interests are not presently welcomed in this world. Pressure mounted from the American Medical Association and pharmaceutical interests and Rife was left abandoned by fearful colleagues.[45]

In **Playing GOD** I've discussed that the universe and all within it has an oscillation rate. We've established that we each have unique signature fields. These signature frequencies (waveforms) are markers of the state of our health –

- The shape of the waveform is important. Jagged incoherent waveforms may indicate disease.

- The frequency of the waveform is important. Rife's work validated micro-organism oscillation rates (frequencies). It demonstrated that frequency alone can be also be used as an indicator of disease.

A simple proof of cellular oscillation is to consider body temperature of a living body versus that of a dead body. When cellular oscillations maintain our body–mind equilibrium we are at body temperature around 37 degrees Celsius. At death the cells cease oscillating and our body temperature becomes that of the surrounding environment.

LAKHOVSKY AND THE DANGER OF MICROWAVES

In the 1920s and 30s author and inventor Georges Lakhovsky experimented with electromagnetic waves to restore cellular equilibrium. This led to him expressing concern about the dangers of microwaves.

Lahkovsky postulated the electrical attributes of plants, humans, bacteria and all life. He thought his Multiple Wave Oscillator would offer a vast spectrum of frequencies from which cells could select the frequency they needed to restore equilibrium. Lakhovsky said, 'I built an oscillator with all the ... frequencies from 750,000 per second to 3 billion. But each circuit also emits many harmonics, which, with their basic waves, their interferences and their effluvia can reach the scale of the infra-red and even that of visible light (1 to 300 trillion vibrations per second).' Cells can 'therefore find on the scale of such an oscillator, the frequencies which cause them to vibrate in resonance.'[46]

This is a similar approach to that of a sound healer working with multiple harmonic frequencies in a sound bath. We are immersed in the multiple harmonics of a gong and our body–mind resonates with certain frequencies. Instead Lakhovsky was utilising electromagnetic frequencies from 750 kHz to 3 GHz and due to harmonics and interference even higher frequencies. (Refer to Appendix D and Figure 14 for more on harmonics.) They are relevant because even though a telco signal emits at say 900 MHz the harmonics and sub-harmonics produced are higher and lower frequencies. This increases the number of radiation frequencies our body–mind must process.

Lakhovsky understood what he was doing was potentially dangerous and that chromosomes and chondromes within the cell could overheat with microwave exposure. 'You can see how dangerous the arbitrary use of short waves with thermal effect can be. This therapeutic method should only be used by practitioners who have a sufficient knowledge of modern physics and biology, for, the majority of physicians who handle these apparatus know very little of both.'[47]

Lakhovsky had reverence for the power of the energies he was working with. Medical practitioners and industry do not have sufficient knowledge, reverence or purity of intention to 'handle these apparatus'. The current apparatus is no longer confined to an enclosed laboratory as in Lakhovsky's time. It includes antenna towers, mobile phones and Wi-Fi emitting microwaves that Lakhovsky noted can overheat chromosomes and chondromes within cells 'like the filament of an overcharged incandescent bulb'.[48]

CHAPTER 19

My Journey V – Salvado por el Movil (Saved by the Mobile Phone)

0 *Acknowledging that mobile phones have their usefulness*

ONCE UPON A TIME I HAD A work project come up in South America. With four weeks before commencement I decided to snowboard the slopes in Argentine Patagonia. Some Spanish first. I stayed in the world's most carnivorous city, Buenos Aires, and started lessons with Nora and Ariel. An English friend, David, had moved from the UK to marry his sweetheart Susana. We'd go out for dinner most nights with Nora and Ariel often joining us. After two weeks of lessons I flew out to Bariloche for early September powder days on the slopes. Lazy café life in Buenos Aires was insufficient preparation for three straight days of snowboarding. My legs tired and the following day I planned down time. Word was the temperature was dropping and the wind was turning gale force for winter's last big storm.

The next morning I caught a bus to Villa la Angostura. The roads were empty as the snow began to build and conditions deteriorated. The bus only just made it through. My destination was Bosque de Arrayanes (Myrtle Forest) and I set off just before midday on a trail I thought was a return loop on the peninsula. With spectacular scenery and no-one around I was in my element. Dark clouds filled the sky and light snow created a spectacular silent ambience. A couple of trekkers coming the other way looked at the sky then looked at me and mumbled, 'Extranjero loco' (crazy foreigner). Snow covered my jeans around the calves. The rest of me was warmed by a Gore-Tex jacket and cheap waterproof gloves. It suddenly started to snow and blow. The trail disappeared under snow cover. I became enraptured. It took 23 years of life before I saw my first snow and every time I was in it I felt like I was catching up on lost time.

Vistas were of a lake and mountains and a backdrop of storm clouds so dark that daytime looked like a moonlit night. By 3.30 pm I sensed the trail was not looping back to the entry point. If I turned back now I wouldn't be back until 7 pm. By 4.30 pm darkness had closed like a black ice curtain. The trail had gone inland and I no longer had the lake as a navigation reference. Landmarks were non-existent. I was knee deep in snow. The storm was as immense as any I'd experienced. By 6 pm early stages of hypothermia were setting in. Due to the snow offering a whisker of sky-light reflection I could see the multi-trunked old-growth myrtle trees straight out of an animated adventure movie. I was without a torch and this whisker of light came with Grace.

By 8 pm I knew I was in for a long night. There was no saving face on this one. I reached into my pocket for my mobile and there was a signal! I called Susana. She'd been to the area and knew the forest. She said she would call Villa la Angostura Emergency Services. They called me and asked where I was. 'Donde estas, Benjamin?' I don't have any idea. 'No tengo ni idea.' Somewhere in the forest. It all looked the same. Translated through Susana they said the weather was too severe for helicopters and I had to find the jetty. When the weather calmed they'd send a boat.

By 10 pm I was hypothermic, shivering uncontrollably and my mental faculties were waning. Snow-wet jeans heightened the wind chill. My mission in icy darkness was to find the jetty. The tiny log cabin bumped into me. It was locked so I smashed the window and wriggled through onto a bench and pitch black darkness. I felt my way around and my eyes adjusted. There was no food and no blankets but at least I was out of the wind. I discovered a small jar of dark sweet-smelling liquid. I scooped snow off my jacket and placed a dollop in the quarter jar of raspberry cordial. Sipping that sugary, preservative-loaded mix soothed my mental state. I called Susana and told her I'd found a cabin. She said the emergency services rescue vessel was on its way and 'Find the jetty, Benjamin!' Docking and a land search was not an option.

I later found out it was minus 14 degrees Celsius and far colder in the wind chill. My options were staying in the cabin and not waking or going back into the elements. I set off through the exquisitely twisted myrtle forest. The first clue was what smelt like a dumping area. There were oil drums and an oily tarpaulin. I wrapped the tarp around me and thought I could vaguely hear waves crashing. The storm clouds parted briefly for skylight to shine through. A second of illumination confirmed there in the distance was the jetty! By now it was after midnight and the boat should have been in the vicinity. I waited and shivered and waited. The phone had been switched off for the last hour. Susana and the rescue team knew it was a lifeline running on empty and had held off calling. I huddled in a shivering ball leaning against one of the jetty pylons.

The tarp cut the wind chill. I started to drift off not into sleep but some other place. As my eyes half closed I glimpsed the blurred lights of a vessel travelling away from the jetty in the snow storm. Had I missed my chance? I leapt to my feet and called out as the vessel moved further away. The storm was so fierce I couldn't hear my own shouts. Waves splashed against the jetty and the oily tarp I'd wrapped myself in. The vessel's lights bobbed up and down. This triggered an impulse to reach into my pocket and open my mobile phone. I yelled and waved it above my head.

The light and the phone died and I slumped dejected. I felt that was my last shot. Then I noticed a different aspect to the bobbing lights. The boat was slowly turning around! I kept waving my arms in the darkness. The rescuers courageously navigated the boat through the snow blanket. In the blissful illumination of the vessel's spotlights two rescuers leapt onto the jetty and I fell exhausted into their arms. The rescuers covered me in woollen blankets, checked my extremities for frostbite and we chugged home briskly to Villa la Angostura. Salvado por el movil. Saved by the mobile phone.

After a 2 am hospital visit confirming severe hypothermia but otherwise okay, the chief rescuer invited me back to Villa la Angostura police station with the rescue team. There was a young police officer and a lady police officer, both in great spirits. Like concerned family members they were curious about my adventure. 'What were you doing out there?' 'Trekeando senora!' 'Just trekking!' They were a loving bunch of humans amongst many loving forces with me that early morning. I swigged hot chocolate and soon dozed off in the police station.

I woke refreshed and to a new crowd at morning roster change. They asked how I was and I replied that I'd slept better than I'd had in years. In a police station. Who would have imagined! I went off and made payphone calls home. I then bought enough artisan chocolates to fatten the Argentinian World Cup team and presented them to the police and rescue team with a card, shed a tear and said my goodbyes. That day my story was in the local paper and Argentina's national paper EL PAIS with the caption 'Salvado Por El Movil'. It had been the worst September weather in the region since weather statistics began. I've been to both extremes. Ten years ago I was saved by a mobile phone and recently I was sickened by one.

CHAPTER 20

An International Focus

- *Comparing radiation pollution policy around the world*
- *The French law to ban Wi-Fi in preschools*

INTERNATIONAL POLICY COMPARISON

AS MY RADIATION POLLUTION SICKNESS LINGERED I pondered escaping to an overseas destination that had lower exposure limits. These daydreams provided solace from my reduced-freedom reality. The vastness of Australia means we can drive for three hours and be in the middle of the boondocks and kilometres from nearest towers but devoid of the lifestyle that we love, friends, family and community. I lived in seven countries for extended periods and enjoyed all of them. Was it time to go overseas again this time as a radiation refugee? Where would be a suitable location? Indonesia and the island of Bali was crossed off the list. It had a population density comparable to inner city Sydney and poorly regulated tower construction. Was there a hut on an East Asian mountain where I could live for a dollar a day?

Mount Everest would not be the place to build a hut. Not only does it cost well over a dollar a day but it now has 4G coverage at 5,200 m above sea level with thanks to Huawei and China Mobile.[1] I was fortunate to spend a few weeks in Mongolia during the northern hemisphere summer 15 years ago. The world's least densely populated country is now 'connected' thanks to the World Bank's International Development Association and Global Partnership of Output-Based Aid. There are now 'Mobile base stations to provide mobile phone services to the population of

90 soum centres (districts) and the surrounding herder areas'.[2]

There are advantages in the herders being better connected. What is the trade-off for the herders? There has been an unprecedented mining boom in Mongolia. The once pristine landscape is in many areas now deformed. Nomadic herders now traverse networks of dusty mining roads and there are more and more stories of herders not being able to water their livestock due to mining project water consumption.[3] Once commodity prices drop and the boom eases the herders are left with watering issues, dust and the withdrawal of international aid. So East Asia was crossed off the list. What if I went further west into Europe?

Scandinavian countries were ahead of us in their uptake of mobile phones. Swathes of Europeans experience the debilitation of radiation pollution sickness. In Sweden radiation pollution sickness or electrohypersensitivity has been a recognised disability since 1995. An open letter to the Swedish IT Minister cites a National Board of Health Environmental Health Report 2009 in which 300,000 Swedes indicated they are detrimentally affected by electromagnetic radiation.[4]

Those affected by radiation pollution in Sweden have the right to employer support so they can maintain work duties. Low emission computers and removal of wireless DECT phones are examples of office alterations to enable employees to continue working. Olle Johansson from the Swedish medical university Karolinska Institute along with seven scientists from five countries formed The Seletun Scientific Statement.[5] They urge vital action to implement new biologically based exposure standards worldwide given the body of evidence.

The Seletun Scientific Statement 10 Key Points are –

1 The global population is at risk
2 Sensitive populations are currently vulnerable
3 Government actions are warranted now based on evidence of serious disruption to biological systems
4 The burden of proof for the safety of radiation-emitting technologies should fall on producers and providers not consumers
5 EMF exposures should be reduced in advance of complete understanding of mechanisms of action
6 The current accepted measure of radiation risk—the specific absorption rate

('SAR')—is inadequate, and misguides on safety and risk

7. An international disease registry is needed to track time trends of illnesses to correlate illnesses with exposures
8. Pre-market health testing and safety demonstration of all radiation-emitting technologies
9. Parity needed for occupational exposure standards
10. Functional impairment designation for persons with electrohypersensitivity[6]

The European Economic and Social Committee, 'a bridge between Europe and organised civil society', convened for a public hearing on 4 November 2014 to discuss electromagnetic hypersensitivity in Europe. The Committee said 'a growing number of Europeans and according to new estimates, between 3% and 5% of the population are electro-sensitive'.[7] Another study indicates the percentages may be even higher (Appendix C). In the Draft Opinion the European Union actively acknowledges the growing number of people with electromagnetic hypersensitivity. It recommends strategies to assist those experiencing it and to prevent more people becoming electromagnetic hypersensitive. A key point is the implementation of ALARA (As Low as Reasonably Achievable) with regard radiation emissions.[8] Implementing ALARA in Australia would mean an 'achievable' factor of 100 reduction in exposure limits by 2017 and a factor of 1000 by 2020. Why a factor of 100?

For microwaves at 2,100 MHz general public exposure in the following countries has an equivalent plain wave power intensity limit of 0.1 W/m² –

- Bulgaria
- Italy – near new homes, in schools and playgrounds and anywhere someone would spend more than four hours
- Lithuania
- Poland
- Russia

Australia's exposure limits for 2100 MHz microwaves are 100 times higher at 10 W/m².[9] A factor of 100 reduction in exposure limits would match Russia and requires little action from government agencies and telcos. It would be a stepping stone to further emission reductions and rewiring in Australia. Device-using consumers would continue using. After telcos have turned down the volume on some of their high emission antennas and Wi-Fi device output is limited we get to reclaim Australian streets. We would be affirming the health and harmony of Australians as the nation's number one value.

Despite the exponential growth of microwave sources and emissions ARPANSA's efforts remain directed towards ionising radiation such as nuclear sources. How about the 2015 version of nuclear that comes from towers and tall buildings? More relevant to Australians is their illness soon after a telco tower was constructed in the neighbourhood. More relevant to Australians is the Wi-Fi in their child's school.

What are Australian researchers doing to contribute to national health? According to their website (at 19 March 2015) our $5 million government funded university researchers on the topic had travelled overseas to conferences but had not published a single paper.[10] Research funds were granted in 2012/2013.[11]

THE FRENCH SAY AU REVOIR TO WI-FI IN CLASSROOMS

In 1996 a group of CSIRO radio astronomers patented wireless local area network (WLAN) technology. This went on to be used in Wi-Fi hotspots and installed on over 5 billion devices worldwide.[12] We are still using this technology two decades later. The microwaves are of higher energy and the devices are smaller but the paradigm of emitting through the universal ether to transmit communications data remains unchanged. It is an unhealthy paradigm that won't transform unless we find another way to communicate and enhance life simultaneously.

In 2010 the NSW Minister for Education and Training helped to form at the time the 'world's largest Wi-Fi network'. It was to cover every school in the state.[13] Even with the technology spend from 2003 to 2012 'Australian teenagers' reading and maths skills have fallen so far in a decade that nearly half lack basic maths skills and a third are practically illiterate'.[14] The 'more money more technology' paradigm doesn't work yet we stay with it.

Dr Marie-Therese Gibson was principal of Tangara School for Girls for 19 years. She resigned in 2013 due to health problems she attributes to the Wi-Fi installed in 2010. 'I realised as time went on I was getting sicker and sicker and couldn't sleep at night. There were parts of the school I just couldn't go into. I started getting strange headaches and tremendous fatigue, and I found I couldn't think clearly. My thyroid is kaput and my body can't make melatonin.'[15]

If screen-based teaching is linked to improved learning outcomes why don't we hardwire? On 29 January 2015 the French National Assembly passed a law to ban Wi-Fi in preschools (under 3) and minimise it for children under 11. Other actions include Wi-Fi hotspot labelling and SAR labelling on mobile phones.[16]

The Australian Medical Association president said '... if we can minimise exposure ... it's probably a good idea'.[17] Despite this children and teachers are exposed to Wi-Fi for over six hours a day five days a week. EMF Essentials talks are designed to inform parents and teachers such that they are empowered to take appropriate action - visit dharamhouse.com

It is time for action. A 2011 Council of Europe Parliamentary Assembly Report indicates, 'It is certain that one cause of public anxiety and mistrust of the communication efforts of official safety agencies and governments lies in the fact that a number of past health crises or scandals, such those involving asbestos, contaminated blood, PCBs or dioxins, lead, tobacco smoking and more recently H1N1 flu, were able to happen despite the work or even with the complicity of national or international agencies nominally responsible for environmental or health safety.'

It concludes '... there is sufficient evidence of potentially harmful effects of electromagnetic fields on fauna, flora and human health to react and to guard against potentially serious environmental and health hazards ... According to the EEA (European Environment Agency), there are sufficient signs or levels of scientific evidence of harmful biological effects to invoke the application of the precautionary principle and of effective, urgent preventive measures.'[18]

CHAPTER 21

Animals and Plants

- *Plants and animals are susceptible to the effects of electromagnetic radiation*
- *How synthetic layers of electromagnetic radiation may disturb navigation*

MODERN POISON

Misty and her father are parked in the family car waiting for her mother. As Misty glances across the road her curiosity peaks.
Misty – 'Why are those trees dead, Daddy?'
Daddy, clueless, searches for an authoritative answer – 'Council workers have poisoned the trees.'
Misty scrunching up her nose – 'Why?'
Daddy – 'So the trees die and they can pull them out and build shops.'
Misty – 'What's poison?'
Daddy – 'Poison is something that is very bad for living things. We don't want to touch it, breathe it, drink it or be near it. Look at the kids playing on the swing.'
Misty doesn't look – 'How do Council workers poison trees?'
Daddy happy to be asked a question he knows the answer to – 'They drill holes into the roots and into the trunk of the tree and pour in a glyphosphate-based chemical such as Roundup.'
Misty's body tenses as she hears this. She doesn't consciously understand it all but her energy field contracts. She looks out the window to the kids playing on the swing noticing that the swing is on the edge of a brown-tinged 'dead zone' – 'Do council workers poison kids too?'

Councils still spray all sorts of chemicals in areas where children play. There is another poison on the block in 2015. Microwaves can sterilise soil in a way that is currently accomplished by highly toxic chemicals. They may be effective at sterilising other life including humans. The OXFORD DICTIONARY defines sterile as 'free from living micro-organisms'.[19]

RURAL TOWERS

It is not just city folk receiving a sterilising dose of microwaves. I was down at the Fleurieu Peninsula south of Adelaide to take photos and measure the output of rural towers. Microwave towers were found on hilltops with cows grazing in nearby paddocks. With the sparse population distribution in rural areas telcos turn up the microwave signal volume of their antennas. People living in a home close to one of these rural towers may be exposed as intensely as someone in a city suburb.

A study showed reduction of milk yields, increased health problems and previously unseen behavioural responses in cows near a TV and radio transmitting antenna. There was a normalisation of behavioural abnormalities within five days when the cows were removed to a different area. The veterinary surgeon author said results pointed to electromagnetic tension as the cause for the abnormalities in proximity to the transmission tower. To develop a thorough causal relationship more work was needed.[20]

A two-month study of mobile phone tower effects on common frog tadpoles placed 140 m from a mobile antenna resulted in the exposed group showing 'low coordination of movements, an asynchronous growth, resulting in both big and small tadpoles, and a high mortality (90%).' The control group was shielded from the electromagnetic radiation inside a Faraday cage. 'The coordination of movements was normal, the development was synchronous, and a mortality of 4.2% was obtained.'[21]

If electromagnetic radiation includes a magnetic component are magnetic fields generated in animals and humans being altered? What happens to our inner compass? Our capacity to perceive direction, location and altitude is taken for granted and given less emphasis due to our city based lives. We no longer need it to survive. In the Patagonian myrtle forest at minus 14 degrees Celsius I needed to navigate to the jetty in order to make it through.

In the Sydney EMF Experiment I attributed stepping off at the wrong train station to brain fog. There were also minor driving direction complications which were atypical. Perhaps magnetoreception or my sense to detect magnetic fields to navigate played a part. 'In vertebrates, such as migratory birds and sea turtles, the ability to sense the Earth's magnetic field is clearly important for positional and directional information during their long-distance migrations.'[22]

Magnetic fields are potentially shifted by layering our world with synthetic electromagnetic radiation. The magnetic field measured below a high voltage power line is different from the Earth's magnetic field. Are migratory birds detecting the Earth's magnetic field or the stacked synthetic magnetic fields of modern society?

Are we detecting the Earth's magnetic field or the stacked synthetic magnetic fields of technology? Our brain contains the mineral biogenic magnetite Fe_3O_4, which is an iron oxide.[23] It was found in the region of the sphenoid and ethmoid bones behind the nasal cavity in front of the pituitary gland.[24] Through magnetic activation and biological coupling regions of our brain proximal to our master glands may be altered through everyday exposure to electromagnetic radiation technologies.

CHAPTER 22

Kool for Kids

- *Smart kids with smart phones*
- *Parental anxiety - keeping up with the Joneses in a microwaved world*
- *Increased radiation absorption in a child's brain*
- *Going through life without speaking face to face with another human*

IN AN INTERESTING INTERMINGLING WITH TECHNOLOGY KIDS are getting smarter. A mobile device is like an extension to a child's brain. They might not be able to do a calculus problem but they are resourceful enough to start a tech company. They may be semi-literate by old educational standards but with savvy multimedia sharing they can conjure up a social media following. I've been startled by the wisdom children offer. Meanwhile another group of kids has lost interest.

Physical device distraction, social pressure to use and the direct energetic disturbance from microwave emissions may all contribute. Wearables are being marketed to children as young as five. At the same time parents are subliminally messaged and consumed with anxiety around the safety of our children. We fear not being a 'good enough' parent. The media feeds this fear and marketers ride it to sell products. If we do not purchase the products and something goes wrong then we will never forgive ourselves. So we wire our lives and our kids so if anything were to happen we can say to ourselves 'I bought the device and did my best.' Samsung have an ad running where a daughter and father are at the beach together with a surfboard. The father captures a photo with a dripping wet phone. The subliminal is that if father didn't seize the moment using a Samsung mobile phone he would be an inadequate father for the rest of his days.

During my relatively loose upbringing, at four years old I rode my bicycle to a friend's house and we traversed town to go fishing. We sneaked out the window at night to go on adventures. Those friends are today's parents frantic to keep their kids in sight. Smart devices are marketed as safety devices for kids. Is the world safer for a kid with a smart phone or wearable in his pocket? I haven't seen a study that indicates it is. We buy it anyway. 'I'd better get the smart wristwatch for Julie ... everyone is getting them for their kids and if anything happened ...'

Lynchburg is the appropriately named US city manufacturing the FiLIP wearable smart locator and phone for kids. They have Forbes, Fox, USA Today, PC Magazine and CNN testimonials. Next Generation FiLIP 2 'as seen on Disney Channel' comes in Superhero Blue, Limesicle Green, Awesome Orange and Watermelon Red.[25] Like healthy fruit juices. The wearable has a Smart Locator to locate our 'child using a blend of GPS, GSM and Wi-Fi.' There is a red Intelligent Emergency button to keep our 'child safe at a moment's notice'.[26] Safe Zone settings notify us when our child leaves that zone.[27]

FiLIP Technology have combined GPS, GSM cell tower location and Wi-Fi hotspot triangulation to provide the power, accuracy and quality of a smartphone in a colourful wristwatch.[28] They target the 5 to 11 year old demographic. My first analogue wristwatch was a hand-me-down gift from my grandfather. I was so excited with it I never took it off. Kids want to be Superhero Blue activated 24 hours a day and it is likely the parents have not been made aware of attributable adverse biological effects of wearable-emitted microwave radiation. There are now multiple wearable options for five-year-olds. But what options exist for the ultra-young? How do mobile manufacturers cater to infants?

Fried young genitals anyone? Parents feeling inadequate will love this one. Wireless nappies may be the next big thing in wearables. 'A disposable organic sensor that can be embedded in a diaper and wirelessly let a carer know it needs changing.'[29]

For 24-hour baby monitoring we have the World's Smartest Dummy. 'The world's first smart pacifier that monitors a baby's temperature and transmits the data to an app on a parent's iOS or Android smartphone or tablet.' The device uses Bluetooth 4.0 technology.[30]

These are all part of a trend towards Orwellian themes and transhumanism. Why don't we just get on with it and embed a 24/7 radiation emitting smart chip in every newborn?

REGULATING DEVICES FOR CHILDREN

ARPANSA advises parents to 'encourage their children to limit their exposure'. They say research relating to children and electromagnetic radiation is limited 'and the possibility of harm cannot be completely ruled out'. ARPANSA indicates they do not know at this time if –

- the exposure received by a child's brain is higher than that received by an adult
- children are more sensitive due to their developing nervous system
- children are more vulnerable due to a longer potential lifetime of exposure.[31]

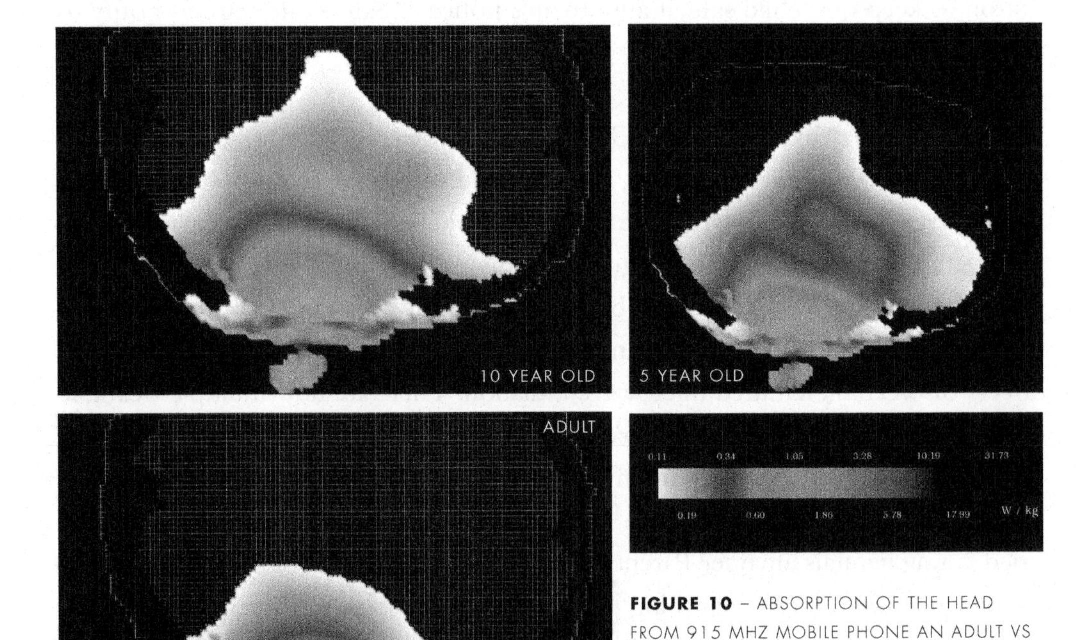

FIGURE 10 – ABSORPTION OF THE HEAD FROM 915 MHZ MOBILE PHONE AN ADULT VS CHILDREN OF 5 AND 10 YEARS[34]

Specific absorption rate (SAR) for a 10-year-old is 153% higher than that for an adult.[32] Gandhi et al conclude that the fluid used in the standardised mannequin for SAR testing has average electrical properties of the head which does not distinguish for varying tissue properties. The lack of anatomical accuracy coupled with disparity in head size between test mannequins and people leads to inadequate SAR testing. SARs are larger for the smaller models of children and there is larger in-depth penetration of absorbed energy in the smaller models.[33] Figure 10 shows increased mobile phone radiation absorption in the head of a 5-year-old compared to a 10-year-old and a 10-year-old absorbing far more radiation than an adult.

SOCIAL NORMS AND TECHNOLOGY

My friend Lisa does not answer her phone. She listens to her voicemail and responds via text or email. Very occasionally she calls to leave a voicemail if she is certain the other party will not pick up. She leaves a voicemail for her husband while he is at soccer practice or for her friend when she knows she has dancing class. She does not risk conversation. However, Lisa leaves substantial voicemails. One I received went for two minutes and included enthusiastic vocal inflection.

We can now go through life having our survival needs met, like Lisa, exchanging requisite information without ever speaking to another human. Why risk talking with someone who might be in a bad mood, rude or who never stops talking? Mobile devices facilitate a light disposable version of connectedness. The slow cooked presence of deep connection is redundant in a fast food world.

When Facebook came on the scene I'd been overseas for many years and was excited to reconnect with long lost buddies around the globe. I had my few hundred friends and there'd be a little ego boost when an upload received a like. On social media everyone was doing shiny things. I didn't remember these people being as superhuman in person. Had their lives really become that wonderful? As I posted the shiny bits of my life I realised the answer was no. My brief foray into social media was enough to show me I felt lonelier and less connected when 'microwave connected.' Technology was a feeble substitute for personal contact and the grittiness of human interaction.

My perspective emanates from having my first exposure to the internet and mobile devices in my late teens. For those born in the last 20 years social devices have dominated communications and they do not know another paradigm. Therefore there is no sense of 'missing' the vulnerability of real connection and presence. Kids I've spoken with long for deeper connection as we all do. They say technology at least gives them some connection. These kids and young adults are perhaps the group most challenged by the concept of switching off for a day or a week as we do on Mobile Free Day (mobilefreeday.org). Those of us 30 or older have had years mobile connected and years mobile free. We know we are not going to 'die' by switching off. We may even be encouraged to switch off by our recollection of profound face to face pre-mobile friendships.

CHAPTER 23

My Journey VI – Run Over By a Mobile Addict

0 *The quadrupled chances of crashing our car when we use a mobile phone whilst driving*

THERE IS SOMETHING ABOUT THE RHYTHM OF driving where we drop our façade and get real with the person next to us. I've shared things I didn't anticipate sharing while on a road trip and I've had people open up to a staggering extent. We might as well do something with all that road time right? If I didn't have a passenger then chatting to a friend using a hands-free phone transformed a dull drive into a shared road trip adventure.

The evolution was to texting. On the drive from Brisbane to the Gold Coast I encountered a bottleneck. I reached for my smart phone and sent a couple of texts to organise a meeting later that day. As the traffic jam eased I relaxed into a road trip reverie in the fast lane of the four lane highway. The mobile phone beeped. I anticipated a text confirming a Gold Coast meeting location. The mobile phone had the attention grabbing power of a crying baby. How could I resist the cute little beep? I picked up the phone and glanced at the text. It required scrolling and it was in that clumsy moment that a pile up almost ensued. The red lights of the car in front of me woke me from mobile trance. I skidded to a heart thumping stop bumper to bumper. My brakes had just been renewed. The low performance vintage Mercedes would have ploughed right through. It didn't end there. I'd had cars tailgating me as if we were Formula One racing. The tailgaters had no time to brake but scooted into an empty half lane to the right.

Unscathed but a little shaken I thought I'd learnt my lesson. I resisted reaching for the phone when I heard the beep. The mystical beep had an electromagnetic pull and a week after a near catastrophic accident I was again glancing at my phone. Soon I was back texting. There was a day when I counted one in ten drivers using their phones. I was irate with them. 'Get off your phone, you are a danger to society.' I knew I was projecting and that I was just as much a danger. After my realisation I cut out temptation. My phone was switched off when driving. Friends were initially annoyed with my delayed response to messages and my unavailability. After a couple of weeks of adjustment it was the new norm.

Texting, surfing the internet or talking on the phone while driving is a top five cause of fatalities on NSW roads. It sits next to speeding, fatigue and drink driving. The NSW Centre for Road Safety says mobile device related accidents are under-reported and 'The problem is that people are addicted to their phones.'[1] In a study of 456 drivers in Perth it was found that drivers using a mobile phone were four times as likely to have a crash leading to hospitalisation. Use of a hands-free set did not improve this statistic.[2]

CHAPTER 24

Stealth Bombardment III – The Mind Has No Firewall

- *Are technologies being developed to interfere with human function?*
- *The explosion of electronic devices and links to the epidemic of neurological disease*

LIEUTENANT COLONEL TIMOTHY L THOMAS WAS AN analyst at the Foreign Military Studies Office in Kansas. He'd been Department Head of Soviet Military-Political Affairs at the US Army's Russian Institute in Germany. In the article 'The Mind Has No Firewall' Thomas asks, 'What technologies have been examined by the United States that possess the potential to disrupt the data-processing capabilities of the human organism? ... Microwave weapons, by stimulating the peripheral nervous system, can heat up the body, induce epileptic-like seizures, or cause cardiac arrest. Low-frequency radiation affects the electrical activity of the brain and can cause flu-like symptoms and nausea. Other projects sought to induce or prevent sleep, or to affect the signal from the motor cortex portion of the brain, overriding voluntary muscle movements. The latter are referred to as pulse wave weapons, and the Russian government has reportedly bought over 100,000 copies of the Black Widow version of them.'[3]

Thomas cited a Russian Army major who asserted 'psy' weapons were under development all over the globe. For example, 'A psychotronic generator, which produces a powerful electromagnetic emanation capable of being sent through telephone lines, TV, radio networks, supply pipes, and incandescent lamps.'[4] Thomas' article is explicit regarding the damage microwaves and low frequency radiation invoke. He talks of peripheral nervous system stimulation and inducing epileptic-like seizures. As we saw in Part I electromagnetic hypersensitivity correlates to changes in our nervous system. Research conclusions per Appendix B confirm persistent change in the nervous system after high frequency exposure.

MULTIPLE ENVIRONMENTAL TRIGGERS AND BRAIN DISEASE

In 2013 there were 322,000 Australians with the medical mystery of impaired brain function known as dementia. Around 400,000 are expected to have dementia by 2020 and 900,000 by 2050.[5] Professor Colin Pritchard of Bournemouth University says there is an alarming hidden epidemic of neurological diseases 'influenced by environmental and societal changes'.

Research shows a sharp rise in dementia and other neurological deaths in people under 74 and early onset for people under 55. Pritchard says, 'It is not that we have more old people but rather more old people have more brain disease than ever before.' The period for the rise in brain disease is too short for the cause to be genetic. 'Considering the changes over the last 30 years – the explosion in electronic devices, rises in background non-ionising radiation, PC's, microwaves, TV's, mobile phones; road and air transport up four-fold increasing background petro-chemical pollution; chemical additives to food, etc. There is no one factor rather the likely interaction between all these environmental triggers…'[6]

CHAPTER 25

Deals with Society and Mind Control

- *Mobile manufacturers have replaced cigarettes and alcohol as the new sports sponsors*
- *The technology-induced programmable trance state*
- *The ways that radio frequencies can alter the mind*
- *Modulated frequencies for mind control*
- *Similarities between microwave radiation and drugs on biological systems*
- *Moulding the minds of the masses*

WHEN I WAS GROWING UP IN THE 1980s, cricket, footy and motor racing were high on the priority list. Big Tobacco including Philip Morris, Rothmans and Amatil were the largest sports sponsors in the country in 1980. Richie Benaud and Tony Greig would commentate Benson and Hedges sponsored cricket matches. In 1989 there was the Marlboro-sponsored Adelaide Grand Prix.[7]

Then we saw the demise of cigarette-sponsored sports advertising. Sponsorship transitioned to beers, fizzy drinks, banks and airlines. Professional surfing has recently been wooed by mobile manufacturer money. The Samsung Galaxy Championship Tour provides a World Surf League platform for Samsung to spruik mobile phones and wearables. Samsung spent around US$14 billion on advertising in 2013 – more than Iceland's GDP.[8] Perhaps, like cigarette advertising in sport, mobile device and telco advertisements will soon be restricted and they will be required to disclose adverse biological effects on their products as in Figure 11.

FIGURE 11 – MOBILE PHONE PLAIN PACKAGING CONCEPT[9]

'Australia's High Court dismissed the plain tobacco packaging case brought against the Australian government by the world's largest tobacco companies.'[10] A study led by Professor Jane Young from the University of Sydney confirms 'real behaviour change following the introduction of plain packaging'.[11]

In the 1930s a cigarette was 'fresh'. In the 1980s there was a Surgeon General warning 'Cigarette Smoking is Dangerous to Your Health'. Three decades later advertising was limited and the requirement for plain packaging was implemented. The corner shop near where I lived didn't sell guns and ammunition in a box but they always had the slow acting cigarette equivalent. Are we going to allow telcos and mobile manufacturers years of unfettered death and disease before we require that they disclose in their advertising? To a large extent it depends what we watch on TV and at the movies.

BRAINWAVES AND SUGGESTIBILITY

In The Dark Knight Batman film, surveillance systems are portrayed as necessary for the side of 'good' to defeat 'evil'. The subjugating belief being reinforced is 'government surveillance is necessary in order to be safe'. Another theme is the wealthy industrialist 'hero' saving 'helpless' society through his benevolence. The audience is warmed to Bruce Wayne from the BATMAN films or Tony Stark from IRON MAN. In the ultra-programmable state of movie watching our brains rewire to acquiesce to the hyper-successful businessman or business. We believe the founder of a technology company to be a superhero on the side. They pursue agendas such as mass vaccination, sterilisation and genetically modified organism propagation and we do not ask questions.

From 0 to 7 years of age we are in the delta and theta highly programmable state hence Francis Xavier's Jesuit saying, 'Give me a child until he is seven and I will give you the man.' We then spend more of our lives in alpha and the majority as adults in beta per Table 2. Beta is a state of alertness but also the state of anxiety, mind chatter and stress. 24/7 beta is exhausting. We live in a beta world and electromagnetic radiation contributes to the pervasiveness of beta.

BETA	ALERTNESS AND COGNITION	13 TO 30 HZ
ALPHA	RELAXATION AND CREATIVE MOMENTS	8 TO 13 HZ
THETA	MEDITATION AND SUPER LEARNING AND HAS BEEN DESCRIBED AS THE IDEAL STATE FOR RE-PROGRAMMING	4 TO 8 HZ
DELTA	DEEP SLEEP	0 TO 4 HZ
GAMMA	STATE OF CONSCIOUSNESS IN A LONG-TERM MEDITATOR GROUP PER A RECENT STUDY[12]	25 TO 100 HZ

TABLE 2 – BRAINWAVE FREQUENCIES

When we watch TV there is an opiate dump of endorphins as our brains cross over from the critical thinking left hemisphere to the right. We are then less apt to reason or think critically. As we move into alpha and alpha/theta we become ultra-receptive and programmable. In this light trance our brains are susceptible to rewiring. Our smart phones are scrollable TV screens. We switch on the TV at home or catch the newsfeed of the latest disaster on our iPhone or iPad and we subtly reprogram to a stressed and anxious baseline state.

VOICES IN OUR HEAD

I became interested in the aspect of mind alteration during the Sydney EMF Experiment. The inspiration was the altered function of my own mind over a short period and the extreme debilitation of radiation headaches. Many years ago I worked in South America and my desk was next to the coffee machine. Colleagues loved to stop by for a chat as they prepared their coffee. It was an opportunity to practise Spanish with them as I delighted in their local coffee beans. I was an eight-cup-a-day, double-shot coffee purist. The only experience that came close to my radiation headache were the days after quitting coffee cold turkey. The coffee withdrawal headache lasted four days. The radiation headache lasted four months.

The Sydney EMF Experiment had shades of Subproject 68 where a converted basement at Allan Memorial Institute was used as an isolation chamber for subject testing. With CIA funding, 'incarcerated, lobotomized apes were kept for months in total isolation. Rubenstein's radio telemetry techniques were adapted so that radio frequency energy was beamed into the brains of the already crazed animals … By early 1966 the lobotomized apes who had survived faced another experiment. They were bombarded with radar waves to the brain to render them unconscious.'[13]

Jose Delgado sought to electrically reboot the brain of schizophrenics and epileptics. He created the Stimoceiver helmet and implants and operated the subject via remote control. Radio frequencies were sent to primates and then later vulnerable humans in experiments to control their emotions and physical movements. His 1969 book PHYSICAL CONTROL OF THE MIND – TOWARD A PSYCHOCIVILIZED SOCIETY confirms his sentiment at that time that now that humans had civilised and tamed surrounding nature it was time to civilise and tame the inner being.[14]

Allan Frey observed in 1962 that 'Using extremely low average power densities of electromagnetic energy, the perception of sounds was induced in normal and deaf humans. The effect was induced several hundred feet from the antenna the instant the transmitter was turned on, and is a function of carrier frequency and modulation … The RF sound has been described as being a buzz, clicking, hiss, or knocking, depending on several transmitter parameters, i.e. pulse width and pulse-repetition rate. The apparent source of these sounds is localized by the subjects as being within, or immediately behind, the head. The sound always seems to come from within or immediately behind the head, no matter how the subject twists or rotates in the RF field.'[15]

The Australian Radiation Protection and Nuclear Safety Agency (ARPANSA) knows of the capacity of pulsed modulated microwaves to induce voices in our heads. In an agency paper it writes that 'exposure to high power pulsed microwave radiation can produce rapid thermoelastic expansion of soft tissue within the head. The resultant acoustic pressure wave may be perceived as an audible noise that is typically perceived as a buzzing or clicking noise.'[16] The paper is neither authored nor dated. On pulsed radiation ARPANSA says, 'Although little information is available on the relation between biological effects and peak values of pulsed fields, it is suggested that, for frequencies exceeding 10 MHz' the equivalent power flux density (in W/m^2) 'as averaged over the pulse width should not exceed 1000 times the reference levels ...'

The agency says 'although it is not entirely clear why a factor of 1000 was chosen it is certainly a convenient number and seems to have been chosen primarily on this basis'.[17] The microwave auditory effect has been researched as a means of deterrent and for purposes of torture. ARPANSA isn't 'clear' why a pulsed field can spike to 1000 times thermal effect based exposure limits. A SAR of 2 W/kg in Australia is permitted to spike to 2000 W/kg for the General Public and 10,000 W/kg for Occupational.[18] Is it safe? No, but it is a 'convenient number'.

MODULATED FREQUENCIES

The Soviets developed the LIDA machine in the 1950s. They used 40 MHz carrier radio waves modulated to < 10 Hz to match brainwave states of sleep and hypnotic receptivity. (Modulation techniques are discussed in Appendix D.) If a machine can put a human to sleep an exceptional war-time advantage could be gained. The machine can also keep people awake or agitated by shifting the pulse frequency and amplitude. For example, if the setting was 30 Hz to match beta brain activity we would be alert and agitated and have difficulty slowing our brainwaves to sleep. Acting on the central nervous system the device's 'curative' factor is 'speech formulae of a hypnotic suggestion as recorded on a magnetic tape and designed to produce a suggestive effect via channels of the second signal system (intellect, mind, psyche)'.[19]

Adey and Bawin showed in 1982 that calcium efflux from the cortex surface of brains of test animals increased when they were subject to 147 MHz and 450 MHz radiation amplitude modulated to between 6 and 20 Hz.[20] These modulated frequencies disruptive to brain function match high theta to low beta brain wave frequencies per Table 2. Pulsed modulated microwaves with signal frequencies coinciding with brainwave frequencies alter our brain utility. Today's electronics enable pulsed modulation, transmission of multiple waveforms and rapid switching of frequencies. The possibilities of mind control are endless.

MICROWAVE RADIATION AS A DRUG

The 1998 US Army document 'Bioeffects of Selected Nonlethal Weapons' was unclassified in 2006 under the *Freedom of Information Act*. The document is another confirmation of government agency awareness of biological effects. The document talks of microwave radiation as a drug. 'Some investigators are even beginning to describe similarities between microwave irradiation and drugs regarding their effects on biological systems. For example, some suggest that power density and specific absorption rate of microwave irradiation may be thought of as analogous to the concentration of the injection solution and the dosage of a drug.'[21]

The term 'drug' implies something consumed (or absorbed) that has biological effects on humans. We have seen how commonly prescription and street drugs have the side effect of addiction. The helicopters that scour the NSW and Queensland hinterland for pot crops might be better utilised in 2015 thermal sensing hidden microwave towers.

PROPAGANDA

Edward Bernays wrote the book PROPAGANDA in 1928 and with this became the initiator of the public relations industry. Bernays was a protégé of Freud. One of his techniques was to use Freudian tension or anxiety to provoke association with a product. In PROPAGANDA he says, 'The conscious and intelligent manipulation of the organized habits and opinions of the masses is an important element in democratic society. Those who manipulate this unseen mechanism of society constitute an invisible government which is the true ruling power of our country. We are governed, our minds are molded, our tastes formed, our ideas suggested, largely by men we have never heard of. This is a logical result of the way in which our democratic society is organized … human beings must cooperate in this manner if they are to live together as a smoothly functioning society. Our invisible governors are, in many cases, unaware of the identity of their fellow members in the inner cabinet.'

Bernays is prophetic in his portrayal of industry fulfilling the manipulated desires of the public. He says '… it remains a fact that in almost every act of our daily lives, whether in the sphere of politics or business, in our social conduct or our ethical thinking, we are dominated by the relatively small number of persons … who understand the mental processes and social patterns of the masses. It is they who pull the wires which control the public mind, who harness old social forces and contrive new ways to bind and guide the world.'[22]

Telcos, public relations, 'the voice for the mobile telecommunications industry in Australia'[23] and mobile manufacturers share his attitude. To them we are the 'masses' with mouldable minds. If we don't see the manipulation we will have dis-empowering decisions made on our behalf for a lifetime. Knowledge and conscious awareness transforms our perspectives. Despite the attempts at mind control we are empowered to listen to propaganda … or not. We decide on appropriate individual and community action and to ride on a healthier train.

CHAPTER 26

My Journey VII – The Gift of EMR and Going Home to Self

- *Observing how sickness can enable our needs to be met, and learning to let it go*
- *'Fighting' illness versus empowered self-healing*
- *Understanding the stress response*
- *The Trauma Key and letting go of stress and trauma*

CARL JUNG BUILT ON WISDOM PAST TO develop the principle of opposites, the principle of equivalence and the principle of entropy. The principle of opposites says for each conscious or unconscious action of our personality there is an opposite reaction elsewhere in our system.[1] We are rigid and disciplined with our colleagues but flexible and disordered with our family. The consequence of the principle of equivalence is reduced family time when we work extra hours to meet a deadline. The principle of entropy leads to re-established balance. After working late from Monday until Saturday to submit our project before the deadline we spend the entire Sunday with our family.

Energy balances no matter the life circumstance. During the Sydney EMF Experiment my physical and mental health were dismal. Often I was consumed by my 'woeful' circumstances. Universal laws do not rest. There was perfect balance. While health was at a low point I began the Sydney EMF Experiment and was delivered insights to share with others. I started writing **Playing GOD**. I experienced an inner transformation with an intensity far greater than if I'd been healthy and financially secure. I was placed in a position to let go.

LETTING GO OF ILLNESS

Many of us have a limiting defect that gives us a sense of certainty and significance. We are playing mixed doubles tennis with friends and Michael says, 'My bad knee is playing up.' This has happened previously but we are sympathetic. 'Get some ice on it and thanks for risking it to play with us for half an hour Michael.' Suddenly he is the hero receiving accolades and attention. We sympathy call him for the next week. In a distorted way his needs have been met.

I once went on a surfing trip to Indonesia with friends and friends of a friend. We did it in style. One bloke I met on the trip was 'the guy who can never afford it'. There we were in tropical paradise living it up like surf kings for a week and all this chap could think about was how broke he was. He didn't buy a beer because apparently he couldn't afford it. This guy had his victim role playing out around money and always being unlucky and knocked down by life. He had years of training in this role and we were drawn into his web. Every evening one of us would shout him beers to show him we loved and appreciated him as a significant human.

How does radiation pollution sickness enable our needs to be met? There is no way to release something if it is satisfying our needs unless we put something else in place to satisfy those needs. Radiation pollution sickness provided me with certainty. The next day would bring poor health. It was a reason for my not thriving in life and playing small. When I needed a dose of significance I could bring it up with family or friends. I was in the special less than 5% of the population category experiencing extreme symptoms. Being debilitated by radiation pollution sickness validated the mental laziness of negative thinking and the opt-out of holding a grudge against an 'unfair world'.

Grieving for loss of health, relationship, finances and work was an important part of the journey. I found it hard to let go. I closed my fists for long enough that they became too stiff to fully open. Harry drops his Easter showbag multi-coloured lollipop in the sand. He splits his parent's eardrums with wails. Thirty seconds later he is on his dad's shoulders laughing at the blue ribbon cow parade. It took me months before I reached a point where I could laugh at the cows.

We then move beyond grief and into a space of appreciation for our radiation pollution sickness. I expressed my gratitude by writing about the Sydney EMF Experiment and my decline and recovery. Not long after the return flight from Bali I realised I wouldn't change a thing about how things had turned out. It was around that time that my symptoms eased.

To prompt a self-inquiry process you might ask yourself –

- What part of me needs the certainty (safety) of holding on to and not moving on from radiation pollution sickness? How can this need be met in a healthier way?

- What part of me needs the significance derived through sympathy from others for having a 'special' illness? How can this need be met in a healthier way?

- Can I let go of needing to understand why I've been chosen for this? If not, what is stopping me let go? What steps can I take to learn to let go?

- How has radiation pollution sickness improved my life? If it feels like radiation pollution sickness has taken away health and freedom dig a little deeper to find the hidden prosperity. Make a list of gains to see the balance in life.

- We can get fixed in a routine of waking up sick and living like a sick person. What would life be like without radiation pollution sickness?

- What other messages are coming from this experience? Are there layers of held stress, grudges, guilt, shame and traumas to release to free up my energy?

- What am I particularly sensitive to? How is my sensitivity a gift? When will I be okay with being sensitive? How can I use this gift in the world?

FEAR AND STRESS

I lived in northern Arizona for a period. In the US they love a fight. Cage fighting. Republicans versus Democrats. Hollywood blockbusters. The war on terrorism. Beating cancer. Fear propaganda has fingers that reach internationally but the palm of the hand is there in the United States of America.

I'd had some crash landings skiing on slopes beyond my current capacity and was visiting a chiropractor. She recommended a series of neck X-rays. I wasn't a fan of X-rays even at that time. 'Is there a way I can be treated with minimal X-ray exposure?' First she emphasised how minimal the exposure was. 'Minimal relative to Chernobyl but not so minimal relative to nature', I replied. I was immersed in America's 24/7 fight or flight mode and this time I was fighting my chiropractor. She pulled out the red 'fear' card. 'Look, your neck vertebrae are almost fused. If we don't get monthly X-rays we don't know if we are progressing.' There was enough logic in what she said for me to proceed. But really it was the fear infused image of an immovable neck that made me supple to her demands. As soon as I said 'go ahead' she took the opportunity to take X-rays of not only my neck but every aspect of my spine.

We visit a doctor and are told we have six months to live. We do what the authority figure tells us we will do and die within six months. If the doctor had focussed on patients that had healed then we would know healing is possible. If he'd honestly said, 'Outcomes are influenced by the individual's emotional state and her environment' we would be self-empowered to heal. We'd be even more inspired when he gives an example of a patient's journey – 'Frank attributed his recovery to a holistic approach and reconnecting with what he loves in life.'

Alternative healers are often as enmeshed in fear as mainstream medicine. I once visited an alternative healer who told me I should put up my light shield to be protected from 'dangerous' energies. My anxiety levels escalated and I questioned how much healing would occur in that state. Afterwards she turned into a ninja, chopping and clearing the energy between us. Her animated shaking off my energy was similar to the way we shake off a stinger at the beach. I felt her fear of picking up 'dirty' energies being projected. What was wrong with me? Did she see my energy as especially 'dirty'? I asked her about it and she said she was as animated with all her clients.

STRESS RESPONSE

The stress response exists to protect us from threatening predators. We see a wild dog barking and sense it is about to leap for our jugular. Our hypothalamus alarm activates a response leading to stress protein, adrenaline and cortisol release (refer to Appendix B). We go into alarm mode and fight or flight. The wild dog opts to turn and run. Our adrenaline and cortisol levels and heart rate slowly return to baseline.

What if this stress response is activated long term? What if it is not a wild dog but everyday life that is enough to trigger the stress response? We do not return to baseline levels and we have cortisol over-exposure. After months and years of this it becomes our baseline. We know only fight or flight which means we live life shut down. Additional toxins from the outside cannot be processed and detoxified. Our cells do not oxygenate while in stress response mode. We are not able to grow while responding 24/7 to 'threats' and therefore cells shut down and die.

During the Sydney EMF Experiment I shut down and died for four months. Even when I left Sydney and went to a Zero EMF Sanctuary I remained shut down. Though environmental toxins had decreased, residual emotional stress triggered a continuation of the stress response and needed to be collapsed.

Acute exposure to electromagnetic radiation paralleled calamitous life events. These events included loss of relationship, job, home and money, and increased conflict. I dumped adrenaline and cortisol as I experienced fear and traumatic events unfolded. This 'locked-in' the stress and trauma to my system with an intensity proportional to the amount of adrenaline released. I developed an association of stressful and traumatic circumstances with electromagnetic radiation exposure. Months later when in proximity to a mobile tower I would tense up. Part of that was the stress response sending me into fight or flight with the message 'this is not a place to be'. Part of it was my electromagnetic field being affected by incoherent waveforms as discussed in Part IV. The other part was the associative response or memory of stressful and traumatic events related to microwave irradiation.

THE TRAUMA KEY

Alana was experiencing low level anxiety. She'd been on and off the healing path for over a decade. Her life was 'fine' but she wanted to relax and enjoy it more. She had no clear correlation with sensitivity to electromagnetic radiation or chemicals. Alana had locked-in trauma from childhood and early adulthood that we needed to dissolve. She'd been to counsellors and therapists for years but reached a point where she felt talk therapy was only reinforcing the trauma. That is when she found me. Years earlier Alana had been the first on the scene after her father-in-law committed suicide. Despite this dramatic event it was actually her childhood memories of feeling 'unsafe' and 'unprotected' that rated higher on her trauma scale.

Held stress and trauma leaks energy that could otherwise be used for getting on with our mission. These traumas are from yesterday, last year, childhood, infancy, conception and beyond. When I left Sydney with my health at a two out of ten I thought an eight or nine would be possible after recovering in a Zero EMF Sanctuary. Instead it plateaued at five. I didn't feel radiation pollution sick but I didn't feel healthy either. There was residual tension and anxiety.

After weeks in Bali and the long flight home I hit another bottom and my health was a one. At that point I surrendered. I looked inside and asked, 'What am I holding on to?' The answer was 'quite a lot'. There are various ways to collapse our stress and trauma. I added on to existing methods to develop processes in THE TRAUMA KEY. It was the transformational jigsaw piece I had been seeking and it came to me when I surrendered and dropped my insatiable seeking. Once I'd collapsed a lifetime of held trauma and recent locked-in associated responses around technology my health was at an eight. Not many people get to live life with health at an eight. My anxiety has dissolved and I felt like I was living again. There was, however, one more piece in the jigsaw that took me from an eight to a ten and truly living.

CHAPTER 27

Dr Radiation First Aid

- *A holistic approach to recovering from radiation pollution sickness – detoxify, recuperate, reintegrate, action and destiny*

DR RADIATION OR DR RAD IS A mnemonic for radiation pollution sickness first aid steps. Due to the multitude of causative factors a holistic approach is optimal. The focus is on *external* rewiring and exposure reduction and *internal* transformation to Detoxify, Recuperate, Re-integrate, take Action and live your Destiny –

DETOXIFY

Either an acute exposure episode or chronic cumulative electromagnetic radiation exposure has taken you past the tipping point. Your priority is to reduce exposure to electromagnetic radiation sources along with any chemical, environmental and psychic pollution. Ask yourself the following –

- Am I exposed to Wi-Fi emissions at home?
- Why not hardwire my internet connection?
- Could I hardwire at work?
- How can I reduce my mobile device use?
- When my mobile phone is switched on is Data Mode disabled to reduce emissions?
- Do I utilise the speaker function?
- Do I switch off devices whilst in my vehicle?
- Could we (household) switch off from 8 pm through 8 am?
- Could I Skype rather than fly cross-country for work?

- Where are the nearest microwave towers to my home? Are they within 500 m? How strong are the emissions (Environmental EME Report per Part II)?
- Is shielding or relocation a consideration?
- Is my bedroom a Zero EMF Sanctuary?
- If I spend my work day immersed in technology detoxifying for at least seven to eight hours a day is critical. Might I rent out of town and commute?
- What life stress can I let go of?

(More detailed radiation reductions strategies are found in Appendix H.)

A heavy metal detox and candida cleanse may be beneficial. In Ayurveda candida can be linked to a blocked third chakra (solar plexus and kidneys). Heavy metals inside our bodies are radiation attracting as is jewellery worn around the neck. A necklace is proximal to the sensitive thyroid gland and thymus. Underwire bras can go. Reports of cut-off lymph drainage and acupuncture point overstimulation of energy channels add to the fact they are antennas. Do I have amalgam fillings? Replace them. Could my metal alloy prosthetic be bioengineered using a polymer?

A liver and gall bladder cleanse is recommended due to the importance of these organs in detoxification. The Liver and Gall Bladder Miracle Cleanse by Moritz is simple and effective. A note on detox. It was tough when I went from eight cups of coffee a day to zero. Detoxification needs to be gentle. I might have reduced my coffee cups to four per day then two per day the next week and one per day the next week before dropping coffee completely. Slow down the process – especially if elimination organs aren't working optimally.

Is mould an issue where you live? Purchase moisture absorbers and light the fire regularly during the wet season. Until you understand this better could you leave the city for a period? Can you get away for Zero EMF camping on weekends? Are you willing to try homeopathic remedies? Bull kelp or brown kelp homeopathic solutions or vibrational essences can take the initial edge off symptoms. We often have these on hand at EMF Essentials talks – to connect for a talk in your area visit dharamhouse.com

Do you sense that exposed glands such as your thymus, thyroid and parathyroid could do with an energy boost?

The Thymus Exercise – The critical glands are all susceptible to the adverse effects of radiation. The thymus is inside the breastbone higher up than the sternum. Twice daily for one minute wrap the fingertips around the thumb tip as if creating a bird beak with each hand and pour energy from the fingertips into the thymus. Follow intuition. There may be guidance to work on other glands including the thyroid and parathyroid in the lower neck throat region.

RECUPERATE

You might think you have lost your freedom. Anger and frustration shadow emotions of your psyche may be coming to the surface. They arise for dissolution. Ask yourself –

- What am I guided to process?
- How can I make the most of this opportunity to go inward?
- Will I allow myself time out and retreat?
- What are my glands and organs telling me?
- Is it time to let go of trauma, attachment, expectation, limiting beliefs, guilt, shame, judgement and grudges?

Various modalities such as THE TRAUMA KEY can assist (visit dharamhouse.com).

RE-INTEGRATE

THE TRAUMA KEY or other modality work can make you feel lighter. It is important to refill the spaces where stress has been released with your own energy. This is the jigsaw piece to take your health from an 8 to a 10. You raise your vibrations by –

1. Doing activities you love to do. What makes you feel good in a sustainable, fulfilling way? Yoga, sports, loving relationships, working on your mission, meditation and being in nature move your energy and increase your radiance and contented 'glow'.

2. Body-mind awareness. Release enough layers that your light shines. Consciously fill gaps or stagnant areas in your energy field with raised frequency light energy. THE TRAUMA KEY talks more about this – you can start by breathing in white light and breathing out darkness, heaviness and stagnation.

Third Chakra Mini-Practice – For something less physically demanding than stretch pose to activate your third chakra you can place one or both hands on your solar plexus. Bring attention to the area and take three slow breaths. Whenever you feel under assault from psychic, electromagnetic or environmental pollution this is a go-to tool. Rather than a fearful technique of 'protecting' yourself this tool is about expanding your own energy from the powerful energy centre of the solar plexus. With an activated chakra you are able to discern the energies entering your energy field and then it is up to you to decide – Do I want to take on these energies or do I now decide to radiate my own energies? Reinforce your decision. 'I now decide to radiate my own energies.'

TAKE ACTION

The path you've traversed and the learnings have opened new opportunities. What calls? You may be enthusiastic to share your knowledge and new skills with others to help them on their journey. After experiencing acute illness and recovery you will likely tend toward the healing path. Speak your truth. Ask people to switch off their phone. Have mobile free dinner parties. In this way you are helping dismantle dysfunctional paradigms and are part of transforming society.

LIVE YOUR DESTINY

When electromagnetic radiation stresses and internal stresses are eliminated from your life you feel light. You have more energy and space to offer your gifts. Harmony, radiant health, clarity and genius will enter your life.

CHAPTER 28

Switching Off

- *The freedom of being mobile phone-free*
- *Joining with other change-makers to switch off*

A FEW YEARS BACK I HAD A blitzkrieg on a vast swathe of freckle face (polka dot) weeds. After congratulating myself I didn't care to look at that area for the next few months. One afternoon I walked past and saw twice the number of weeds that had been there previously. After pulling the weeds months earlier I had left the empty soil unplanted and nature had filled the void. I spent another day pulling fresh weeds and this time immediately planted the area out with grass seed and veggies. The weeds were mostly crowded out and the substituted grass and plants thrived.

When we reduce or remove our reliance on mobile devices we need to fill the generated void with a healthy substitute. I quit my mobile for a period of five months and the void was planted out with a landline, a $20 payphone card and an already established hardwired internet. I primed friends and relatives by being difficult to reach on my mobile for months before I completely switched off.

During the honeymoon phase it was easy and I had more space in my day. Communication was less frequent but of higher quality in a similar way as when we see family once a year and optimise the time with them. Frustrating moments arose too as I had no GPS to find Dulwich Road or I'd arrive in a town without an internet café so was unable to check email while travelling. Sometimes I had news worthy of a text but too insignificant for a call. These minor frustrations soon faded.

After five months of being mobile device free I chose to commence a minimalist plan. The phone now provides a backup when I'm away from a landline. I leave it switched off 99% of the time and open it to check voice messages and reply to text messages at a cost of less than $10 a month.

I QUIT MY MOBILE – AND YOU CAN TOO

Taking a complete timeout from my mobile showed me I could do it. I found others were interested in doing the same so we put together an opportunity to switch off for a day, a week, a month or forever – together. Keep an eye on www.mobilefreed-ay.org for details. I had more fun and success switching off by following these steps:

1. **Prepare** – A preliminary phase to assist friends, family, clients and colleagues adjust to our new communication style. The warm-up can be switching off from 8 pm to 8 am for two weeks. Then switch off from midday until 8 am. You might open your mobile to check for messages a few times a day if necessary.

2. **Organise and purchase** – I used my phone to navigate so purchasing a GPS SatNav for my vehicle was essential. GPS devices utilise relatively lower intensity high frequency radiation via satellite communications. Or purists can use street directories. The lost time is minimal. A landline at home and work will be an imperative (or hardwired internet phone) and a public payphone card helps if you are travelling.

3. **Inform** – Tell friends, family, clients and colleagues your preferred form of contact which is now landline, email, a regular visit for Sunday lunch … or telepathy.

4. **Share** – Switch off with others and have fun with it! Probably someone in your circle of friends wants to quit their mobile device too. Join others on Mobile Free Day www.mobilefreeday.org, which was created to support a growing contingent of people wanting to switch off together.

5. **Listen** – Spend a moment each day journaling whatever comes up. What are the frustrations of being mobile device free? What are you enjoying about it? How device attached were you? How can your communication be enhanced

without a mobile device? Are your thoughts clearer with the device switched off? What about being mobile device free has surprised you?

6 **Celebrate** – Switching off is brave. Celebrate the achievement with gelato or chocolate cake. Tell others about it. Sharing knowledge and awareness empowers others and enables the next layer of wisdom to flow through you on your journey.

PEOPLE UNITE – THE NORTH COAST EXAMPLE

When coal seam gas came to town in northern NSW in 2012 it invoked fear. Residents were overwhelmingly (many areas > 95%) in favour of remaining gas free. The trucks marched in. As Yoda said in STAR WARS, 'Fear is the path to the dark side. Fear leads to anger. Anger leads to hate. Hate leads to suffering.'[2] Protesters were suffering.

Around the time of the Doubtful Creek Blockade in early 2013 the language was 'anti-CSG', 'threat', 'blockade' and 'protest'. Big blue and white bumper stickers read 'DON'T WRECK OUR LAND & POISON OUR WATER – NO COAL SEAM GAS'. There was tension in communities. Cafés closed their doors to customers who didn't align to their side of the conflict. Simultaneously the whirlwind Lock the Gate movement led by Drew Hutton spread beyond 'Greenies and hippies' to align with conservative farmers. People who had never previously joined environmental protests were involved. There was still anger and a feeling of disempowerment at the time of the Doubtful Creek Blockade but it was starting to shift.

Musicians came from all over Australia to share their voice and the transformation from fear to harmony had begun. By the time of the Bentley Blockade in 2014 the movement had captured all walks of the community including doctors, mechanics, bankers and sportspersons. Events were joyful co-creative celebrations rather than 'protests'. How could they not be when participants were shoulder to shoulder with the merriment of 'Knitting Nannas'? The transformation to a tone of empowered 'freedom' rather than fearful 'anti' was captured by the shift in bumper stickers to 'Coal Seam Gas Free Northern Rivers' and 'Gasfield Free Communities'.

CSG united communities around shared resources (land and water) and in response to the self-interest of the few decimating the livelihood of many. The tragedy of the commons is mirrored in unabated telco tower construction. Through the shared invisible resource that is the universal ether, telco antennas and mobile and wireless technologies debilitate 3–5% of the population. 35% may experience mild to moderate health symptoms. Those not in these categories are experiencing DNA damage despite the lack of obvious symptoms. Could we use the Lock the Gate CSG model to activate a co-creative movement where we request more creativity from industry and government agencies? Might we stand up for the vulnerable and for the value of thriving health for all Australians?

Lock the Gate was a local movement that excited the national sensibility. One region said 'no' and others soon followed. If a community refuses a tower in their suburb they have made it easier for the neighbouring suburb to say 'no'. When a local school hardwires other schools are encouraged to take action to enhance the health of their students and teachers.

A hospital becomes a Zero EMF Sanctuary and other hospitals and aged-care centres begin to switch off. With reduced cellular stress patients have the energy to heal!

MobileFreeDay.org shares awareness and seeks radiation exposure limits 100 times lower than they are today by 2017. We seek levels 1000 times lower by 2020 and 1,000,000 times lower by 2030. Is it achievable? Russia and other European nations are already at our 2017 target level. According to the precautionary principle only health-enhancing mobile telephony technology and modulation would be acceptable. Do we need to rewire? Absolutely – 3G and 4G do not enhance health.

I've taken the Lock the Gate Code of Conduct for Nonviolent Direct Actions (NVDA)[3] and rewritten the phrases to align with co-creative actions for a healthy and harmonious universal ether and humanity –

- I treat each person (including workers, police and media) with respect.
- I connect with people and seek harmonious outcomes.
- I use only civil language.
- I assist the vulnerable when required.
- I accept responsibility for my actions.
- If arrested I remain respectful and courteous.
- I assert my right to speak my truth and my right to silence before the law.
- For humanity I channel anger or despair into constructive emotions.

Conclusion

I ATTENDED A RECENT RETREAT WHICH WAS held two mornings over a weekend. On the first morning I asked if we'd all like to switch devices off, not just volume down but completely off. Most were eager to do this but a few in the group resisted this deviation from their customary practice. During the course of the morning a phone beeped. It didn't bother me, but it irked another participant who was enjoying the relatively microwave-noise-free space.

I offered the cigarette analogy. Would anyone object if I sealed the windows and started chain smoking? I spent 30 seconds explaining microwave radiation. It wasn't the topic of the gathering but people wanted to know more. The next morning the lady who'd been most resistant to turning off reminded everyone to check their phones were off. Overnight she had become a spokesperson for switching off!

Through EMF Essentials talks (visit dharamhouse.com) I am privileged to share insights which shift people's perspectives. Someone came to a talk as a two-hour a day mobile phone user and left self-regulating his phone use to 15 minutes per day. This is individual transformation.

A mother immediately implemented the 8 pm to 8 am technology switch off at home to reduce the emissions her young sons were exposed to. This is family transformation.

We've started democratic processes to decide if a school remains with Wi-Fi or moves to hardwire. People have used the knowledge to dispute tower construction. They've stumped authorities and industry with their empowered questions. This is community and societal transformation.

Another way of co-creating is working with collective consciousness. People are touched through books, an internet blog or video, movies, distance healing and sending a loving thought. The power of loving intention has been applied by yogis for thousands of years and more recently Maharishi Mahesh Yogi. This has been called the 'Field Effect' of consciousness or 'Maharishi Effect'. A study of 11 US cities of population over 25,000 showed with 1% of the population participating in Transcendental Meditation increasing crime rates were reversed.[4]

Later studies on the Global Maharishi Effect showed that when the square root of 1% of the world's population practised transcendental meditation and TM-Sidhi program in an assembly (7000 people at that time or just under 9000 people today) international conflicts decreased by over 30%. This was determined by a content analysis of the NEW YORK TIMES and LONDON TIMES.[5] This healing transcends the space – time constraints we impose upon ourselves. It is healing of past, present and future. The modality does not have to be transcendental meditation. It could be another form of meditation, yoga, group singing or even a joyful gathering of watermelon growers.

For now our journey together is complete. To continue this journey organise an EMF Essentials talk in your local area, business or child's school by visiting dharamhouse.com. For the 3–5% of you who are 'pioneers' experiencing debilitating radiation pollution sickness or the 35% with mild to moderate symptoms, I wish you a DR RAD recovery. For those of you ready to take action to co-create a healthy and abundant society may your path be one of self-realisation, thriving health and harmonious development.

Glossary

Absorption – A wave attenuates or loses energy due to conversion of energy into another form. A mobile phone subjects a user to radio waves that reduce in energy the further into the head they propagate due to heating through transfer of energy.
Blood–brain barrier – A protective barrier that prevents the flow of toxins into sensitive brain tissue
Cellular stress response – Response to environmental stressors such as toxins,temperature, infection, mechanical harm and stress. The response is to protect the cell. Cells even commit suicide (apoptosis) to avoid damaging neighbouring cells that would have otherwise died from environmental stress exposure.
Coherence – When two waves have a constant phase difference and the same frequency they are coherent. Indicates cohesiveness, connection and harmony as opposed to noise.
Conductivity – The property whereby a medium enables electric fields, heat or sound to flow through it.
Cumulative electromagnetic radiation – A created term for the sum of historical accumulation of non-ionising radiation exposure experienced by an individual.
Demand switch – Device to switch off circuits from the switchboard when not in use thereby eliminating electric fields in that room.
Electrosensitivity – Also known as electrohypersensitivity, electromagnetic radiation hypersensitivity and radiation pollution sickness. Occurs when an individual experiences adverse health effects proximal to devices or technology emitting electrical, magnetic and electromagnetic fields.
Electromagnetic radiation – Radiation energy as electromagnetic waves in space.
Electrosmog – Used interchangeably electromagnetic radiation pollution.
Exposure – When electromagnetic fields are experienced by a person at readings higher than those of background and physiological processes then exposure has occurred.
Faraday cage – An enclosure made of conductive material which attenuates electromagnetic radiation entering from the outside. It also amplifies electromagnetic radiation emitting inside the enclosure due to reflectivity.

Frequency – The number of sinusoidal cycles completed by electromagnetic waves in 1 second, expressed in hertz (Hz).

Melatonin – A hormone produced in the brain by the pineal gland and a potent anti-oxidant that protects against free radical and DNA damage.

Microwaves – Electromagnetic waves with wavelengths of approximately 30 cm (1 GHz) to 1 mm (300 GHz).

Milligauss (mG) – A measure of ELF intensity and is abbreviated mG. Used to describe magnetic fields from appliances, power lines, interior electrical wiring.

Microwatt (μW) – Radiofrequency radiation in terms of power density is measured in microwatts per centimeter squared and abbreviated ($\mu W/cm^2$). Used when referring to wireless and tower emissions.

Non-thermal effect – An effect of radiation not related to heating of tissue. It encompasses biological effects and spiritual effects.

Occupational exposure – All exposure to EMF experienced by individuals in the course of performing their work. Limits are higher than limits for the public.

Public exposure – All exposure to EMF experienced by individuals during their day excluding work and medical procedures.

Radiation pollution sickness – Term coined for electrosensitivity or electromagnetic hypersensitivity. The term captures electromagnetic hypersensitivity as an environmental illness rather than a deficit in the person.

Radiofrequency (RF) – The frequencies between 100 kHz and 300 GHz of the electromagnetic spectrum.

Resonance – Tendency of a system to oscillate with greater amplitude at certain frequencies.

Specific Absorption Rate (SAR) – A calculation of how much RF energy is absorbed by the body region per gram of flesh. Measured in watts per kilogram of tissue (W/Kg). The US permits 1.6 W/kg energy into 1 gram of brain tissue from a mobile phone. In Australia and Europe we allow 2 W/kg energy into 10 grams of brain tissue from a mobile phone.

Wi-Fi – Wireless fidelity. Wi-Fi forms zones of wireless radiofrequencies. The zones can range from a household to a large restaurant or CBD depending on emitted power density. The range is typically up to 100 m.

WiMAX – Wireless interoperability for Microwave Access is a telecommunications technology to provide wireless data over long distances typically transmitting up to 10 – 30 km. Therefore the wireless transmitter is very powerful.

Zero EMF Sanctuary – A zone free of human generated electromagnetic radiation as close to background levels as is achievable; also known as a 'white zone'.

Abbreviations

ACEBR – Australian Centre for Electromagnetic Bioeffects Research distributes government funds to academics for radiofrequency research
ACMA – Australian Communications and Media Authority is 'the government body responsible for the regulation of broadcasting, the internet, radio communications and telecommunications'
AMTA – Australian Mobile Telecommunications Association is 'the voice for the mobile telecommunications industry in Australia'
ARPANSA – Australian Radiation Protection and Nuclear Safety Agency is the 'federal government agency charged with responsibility for protecting the health and safety of people, and the environment, from the harmful effects of ionising and non-ionising radiation'
DECT – digital enhanced cordless telephone
DNA – deoxyribonucleic acid contains the genetic instructions for our development and functioning
EHS – electromagnetic hypersensitivity
ELF – extra or extremely low frequency
ELF-EMF – extra or extremely low frequency magnetic field
ELF-MF – extra or extremely low frequency magnetic field
EMF – electromagnetic field
EMR – electromagnetic radiation
FCC – Federal Communications Commission is an independent US government agency to oversee regulation of communications technologies
GSM – Global Systems for Mobile Communications introduced in 1991 and associated technology for 2G
IARC – International Agency for Research on Cancer
IEEE – Institute of Electrical and Electronics Engineers
ICNIRP – International Commission on Non-Ionizing Radiation consists of a main Commission of 14 members, 4 Scientific Standing Committees covering Epidemiology, Biology, Dosimetry and Optical Radiation and a number of consulting experts
kHz – kilohertz
kV – kilovolt
LTE – Long Term Evolution technology or 4G
MF – magnetic field
MHz – megahertz
mT – millitesla
μW – microwatts
mW – milliwatts
mG – milligauss
NBN – National Broadband Network 'is an Australia wide project to upgrade the existing fixed line phone and internet network infrastructure'
RF – radiofrequency radiation
UMTS – Universal Mobile Telecommunications System introduced in 2001 and associated with 3G along with CDMA (Code Division Multiple Access)
Wi-Fi – See Glossary
WiMAX – See Glossary
WHO – World Health Organization

APPENDIX A

Technology Summary in an Alfoil-Covered Nutshell

Electromagnetic fields (EMF) are also referred to as electromagnetic radiation (EMR). The radio frequency segment of the electromagnetic spectrum in Figure 1 encompasses background and location specific electromagnetic radiation sources as shown in Table 3.

BACKGROUND ELECTROMAGNETIC RADIATION	LOCATION SPECIFIC OR 'ON PERSON' RADIATION
*MOBILE PHONE TOWERS (ANTENNAS, BASE STATIONS)	*MOBILE PHONES/IPADS/WEARABLE SMART DEVICES
*NBN FIXED WIRELESS (TOWERS/ANTENNAS, ANTENNAS ON HOMES ADDITIONALLY)	*WI-FI
*WIMAX	TVS, HAIR DRYERS, ELECTRIC BLANKETS
DIGITAL TV	MOTOR VEHICLES
AM/FM RADIO	LAPTOP COMPUTERS
SATELLITE COMMUNICATIONS	LIGHTING INCLUDING LIGHT DIMMERS
POLICE, FIRE, STATE EMERGENCY SERVICES	BLUETOOTH
MILITARY INSTALLATIONS ESP. MISSILE DEFENCE RADAR AND SURVEILLANCE	BABY MONITORS
AIRPORT FLIGHT SURVEILLANCE AND WEATHER MONITORING	INDUCTION STOVE COOKTOPS
AMATEUR RADIO	COMPUTER GAME CONSOLES

BACKGROUND ELECTROMAGNETIC RADIATION	LOCATION SPECIFIC OR 'ON PERSON' RADIATION
GLOBAL POSITIONING SYSTEM (GPS)	SECURITY ALARM SYSTEMS/MOTION DETECTORS
ELECTRIC POWER LINES ESP. HIGH VOLTAGE	ELECTRICAL WIRING IN HOMES

TABLE 3 – SUMMARY OF THE SOURCES OF MAN-MADE ELECTROMAGNETIC RADIATION

*The focus of **Playing GOD** is on the highlighted technologies, which are currently in a growth spurt.

APPENDIX B

BioInitiative Report 2012 Summary

The first BioInitiative Report was released in 2007 concluding that biological effects of electromagnetic radiation were evident at exposure limits well below public exposure limits. The 2012 version updates with contributions from 29 international authors. Ten MDs, 21 PhDs and 3 Masters degreed authors from Sweden (6), USA (10), India (2), Italy (2), Greece (2), Canada (2), Denmark (1), Austria (2), Slovac Republic (1) and Russia (1).

The report assesses scientific evidence on health impacts from electromagnetic radiation at levels below current public exposure limits. Around 1800 studies have been done in the five years between reports. There is detail of adverse biological effects at exposure limits sometimes hundreds of thousands of times lower than public exposure standards.[1] BioInitiative Report Conclusions (condensed by the author) follow in Table 4:

GENOTOXICITY (DNA DAMAGE FROM RFR AND ELF)
Toxicity to the genome can lead to a change in cellular functions, cancer, and cell death. We can conclude that under certain conditions of exposure RFR is genotoxic. Data available are mainly applicable to mobile phone radiation exposure. More studies show effects than do not. Frequency, intensity, exposure duration, and the number of exposure episodes can affect the response, and these factors can interact with each other to produce different consequences. Extremely-low frequency (ELF-EMF) has also been shown to be genotoxic. More studies show effects than do not.

STRESS RESPONSE

Cells react to an EMF as potentially harmful by producing stress proteins (heat shock proteins or hsp).

Direct interaction of ELF and RFR with DNA has been documented and both activate the synthesis of stress proteins.

A wide range of frequencies are active. Field strength and exposure duration thresholds are very low.

Molecular mechanisms at very low energies are plausible links to disease (e.g. effect on electron transfer rates linked to oxidative damage, DNA activation linked to abnormal biosynthesis and mutation). Cells react to an EMF as potentially harmful.

DNA damage (e.g. strand breaks), a cause of cancer, occurs at levels of ELF and RFR that are below the safety limits. There is no protection against cumulative effects stimulated by different parts of the electromagnetic spectrum.

IMMUNE FUNCTION EFFECTS

Human and animal studies report large immunological changes with exposure to environmental levels of EMFs many at levels equivalent to everyday life.

It is possible that chronic provocation by exposure to EMF can lead to immune dysfunction, chronic allergic responses, inflammatory responses and ill health if they occur on a continuing basis over time.

Specific findings: decreased count of T lymphocytes; negative effects on pregnancy (uteroplacental circulatory disturbances and placental dysfunction with possible risks to pregnancy); suppressed or impaired immune function; and inflammatory responses which can ultimately result in cellular, tissue and organ damage.

Electrical hypersensitivity is reported by individuals in the United States, Sweden, Switzerland, Germany, Denmark and many other countries of the world. Estimates range from 3% to 10% of populations, and appears to be a growing condition of ill-health leading to lost work and productivity.

Evidence has been ignored in the current WHO ELF Health Criteria Monograph.

NEUROLOGY AND BEHAVIOURAL EFFECTS

Studies on EEG and brain evoked-potentials in humans exposed to cellular phone radiation predominantly confirm the ability to alter brain wave activity. In behavioural experiments there was evidence of persistent change in the nervous system after exposure to RF (for example days after).

BRAIN TUMOURS AND ACOUSTIC NEUROMAS
Studies on brain tumours and mobile phone use > 10 years give a consistent pattern of an increased risk for acoustic neuroma and glioma, most pronounced for high-grade glioma. The risk is highest for ipsilateral exposure.
LEUKAEMIA
Evidence suggests that childhood leukaemia is associated with exposure to power frequency EMFs either during early life or pregnancy. Up to 80% of childhood leukaemia may be caused by exposure to ELF. Measures are required to keep ELF-EMF from transmission and distribution lines below 1mG. Current Australian Standard for 50/60 Hz allows for 1 G for continuous exposure, could be 24 hours a day, which is 1000 times the BioInitiative recommended. For a few hours per day 10 G is allowed which is 10,000 times the BioInitiative recommended.
MELATONIN, ALZHEIMERS DISEASE AND BREAST CANCER
There is strong epidemiologic evidence that long-term exposure to ELF-MF is a risk factor for Alzheimers disease. Considerable in vitro and animal evidence indicates melatonin protects against Alzheimer's disease. It is very possible that low levels of melatonin production are associated with an increase in the risk of AD. There is sufficient evidence from in vitro and animal studies, from human biomarker studies, from occupational and light-at-night studies, and a single longitudinal study with appropriate collection of urine samples to conclude that high MF exposure may be a risk factor for breast cancer. An association between power-frequency ELF and breast cancer is strongly supported in the scientific literature.

TABLE 4 – BIOINITIATIVE REPORT 2012 CONCLUSIONS (CONDENSED BY THE AUTHOR)[2]

APPENDIX C

Electromagnetic Hypersensitivity by Country

MEASURED YEAR	% ELECTRO-SENSITIVE	COUNTRY	YEAR REPORTED	REFERENCE
1985	0.06	Sweden	1991 (0.025–0.125%)	National Encyclopedia Sw., 1991
1994	0.63	Sweden	1995	Anonymous est., 1994
1995	1.50	Austria	1995	N Leitgeb et al, 1995, 2005
1996	1.50	Sweden	1998	SHBHW, Env. Report, 1998
1997	2.00	Austria	1998	N Leitgeb et al, 1998, 2005
1997	1.50	Sweden	1999	L Hillert et al, 2002
1998	3.20	USA, California	2002	P Levallois, 2002
1999	3.10	Sweden	2001	SHBHW, Env. Report, 2001
2000	3.20	Sweden	2003	Sw Labour Union Sif, 2003
2001	6.00	Germany	2002	E Schroeder, 2002
2002	13.30	Austria	2003 (7.6-19%)	B Spiβ, 2003
2003	8.00	Germany	2003	Infas, 2003
2003	9.00	Sweden	2004	Elöverkänsligas Riksförbund, 2005
2003	5.00	Switzerland	2005	Bern, Medicine Social, 2005
2003	5.00	Ireland	2005	This is London, 2005
2004	11.00	England	2004	E Fox, 2004
2004	9.00	Germany	2005	Infas, 2004

TABLE 5 – PREVALENCE OF ELECTROSENSITIVITY BY COUNTRY, 2006[3]

APPENDIX D

Waveform Physics

FREQUENCY AND WAVELENGTH

Frequency is inversely proportional to wavelength $f = c/\lambda$ where c is the speed of light in a vacuum 299,792.458 km/s, f is the frequency and λ is the wavelength. A telco microwave frequency of 1800 MHz has a wavelength of 16.6 cm which is approximately the ear to ear length of an adult head. Photon energy is directly proportional to frequency. As we move along the electromagnetic spectrum of Figure 1 from power lines to microwaves across to X-rays and nuclear sources the photon energy increases.

UNITS

Power density – In microwatts per centimetre squared ($\mu W/cm^2$) is the measurement of power density for radiofrequency radiation including mobile towers, Wi-Fi and TV/Radio broadcast stations.

Specific absorption rate (SAR) – In watts per kilogram (W/kg) is the measurement of power absorbed by tissue of the body. SAR is used for mobile phone testing. In Australia 2 W/kg is allowed over 10 grams of tissue for an averaging time of 6 min.

Magnetic field intensity (ELF-MF) – Measured in milligauss (mG). Attributed to overhead and buried power lines, lights, vehicles, aeroplanes and appliances.

Electric field intensity (E) – Measured in volts per metre (V/m) and covers the electrical component of power lines and internal house wiring.

IONISING RADIATION VERSUS NON-IONISING RADIATION

Electromagnetic radiation used in communications is non-ionising radiation that does not carry sufficient photonic energy to eject an electron from an atom. Photon energy is less than 10 eV, which is less than required to produce ions by ejection of orbital electrons from atoms.

Ionising radiation has a higher frequency than non-ionising radiation and enough photon energy to eject an electron from an atom. This process is ionisation. X-rays and nuclear radiation are in this category. The Chernobyl and Fukushima reactor meltdowns led to DNA damage from ionising radiation. Research indicates DNA damage occurs similarly from non-ionising electromagnetic radiation.[5]

MODULATION

A message signal of different frequency to the transmission (or carrier) frequency can be achieved through modulation. The carrier waveform is the continuous waveform of high frequency that 'carries' the signal with the information to be transmitted. For example an 1800 MHz carrier frequency may be modulated to deliver a signal of lower frequency of say 100 Hz.

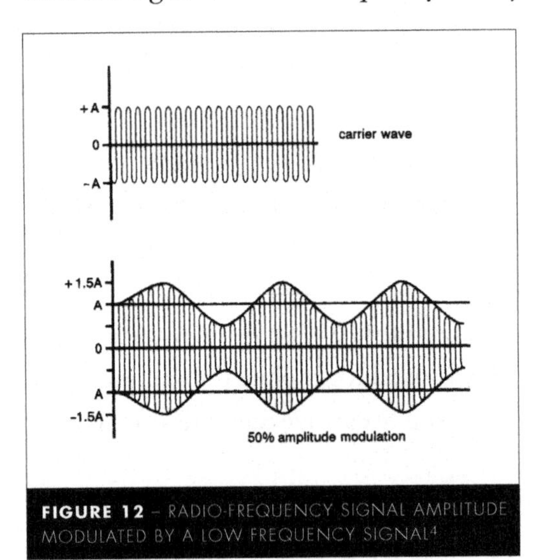

FIGURE 12 – RADIO-FREQUENCY SIGNAL AMPLITUDE MODULATED BY A LOW FREQUENCY SIGNAL[4]

Amplitude modulation – Figure 12 shows an amplitude modulated signal with waveform amplitude at + 1.5 A /– 1.5 A. This could be AM radio. The message signal amplitude in this case is 1.5 times greater than the amplitude of the carrier wave.

Frequency modulation – In FM radio the low frequency signal is embedded as slight changes in the carrier wave frequency.

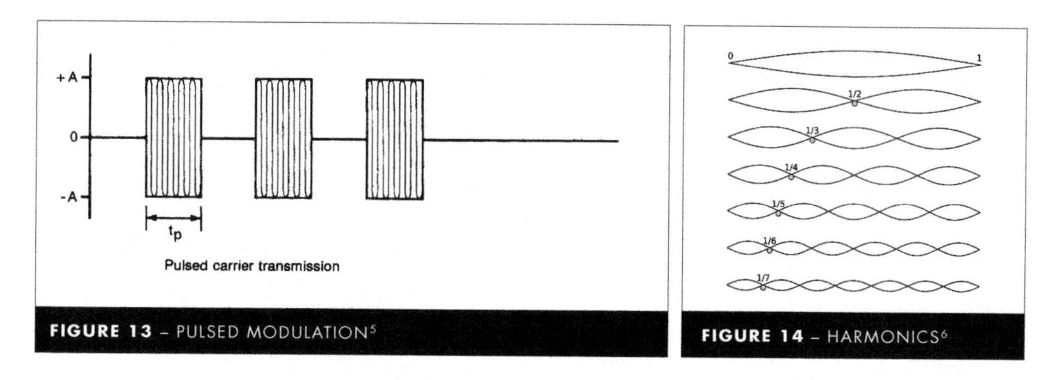

FIGURE 13 – PULSED MODULATION[5]

FIGURE 14 – HARMONICS[6]

Pulsed modulation – Per Figure 13 pulsed modulation uses linear ON/OFF signal switching which is jarring for non-linear electromagnetic humans and Nature. Transmissions are permitted to instantaneously pulse at 1000 times Australia's thermal-based exposure standard. This means in Figure 13 if amplitude A is the standard then an instantaneous spike to 1000 A is permitted.

ON-BODY METER MEASUREMENTS

Directional high frequency radiation metering has the advantage of highlighting the direction of significant radiation contribution. With this information we can distinguish the contribution from each source. For example the tower to the north has twice the impact of the Wi-Fi from the adjoining apartment to the south. Wireless from the apartment below might be measured as a less significant source of radiation.

On-body metering determines exposure independent of direction. It measures the sum of radio frequencies passing through us and is therefore a useful snapshot of total exposure levels at a given moment.

HARMONICS

When a string oscillates as in Figure 14 let's say the fundamental frequency at the top of the figure is 100 Hz. The harmonic at ½ the wavelength or twice the frequency is 200 Hz. The harmonic at 1/3 the wavelength or three times the frequency is 300 Hz and so on. Subharmonics of a fundamental frequency of 100 Hz are at twice the wavelength and ½ the fundamental frequency so 50 Hz. The next one at 25 Hz and so on.

INVERSE SQUARE LAW FOR DISTANCE FROM SOURCE

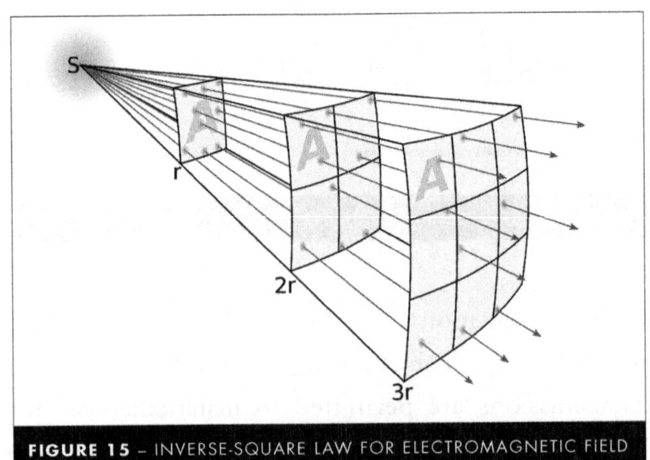

FIGURE 15 – INVERSE-SQUARE LAW FOR ELECTROMAGNETIC FIELD INTENSITY EMANATING FROM SOURCES [7]

The inverse-square law indicates field intensity is inversely proportional to the square of the distance or Intensity α $1/\text{distance}^2$. In Figure 15 say source S is a mobile antenna. If distance r is 1 km then at a distance of 2 km (2r) the radiation intensity is ¼ that at r. At a distance of 3 km (3r) the radiation intensity is 1/9 the intensity at r. The same law applies to Wi-Fi and mobile device sources.

APPENDIX E

International Building Biology Guidelines[8]

Building Biology Evaluation Guidelines for Sleeping Areas SBM-2008, Page 1		No Concern	Slight Concern	Severe Concern	Extreme Concern

A FIELDS, WAVES, RADIATION

1 AC ELECTRIC FIELDS (Low Frequency, ELF/VLF)

		No Concern	Slight Concern	Severe Concern	Extreme Concern
Field strength with ground potential in volt per meter	V/m	< 1	1 - 5	5 - 50	> 50
Body voltage with ground potential in millivolt	mV	< 10	10 - 100	100 - 1000	> 1000
Field strength potential-free in volt per meter	V/m	< 0.3	0.3 - 1.5	1.5 - 10	> 10

Values apply up to and around 50 (60) Hz, higher frequencies and predominant harmonics should be assessed more critically.

ACGIH occupational TLV: 25 000 V/m; DIN/VDE: occupational 20 000 V/m, general 7000 V/m; ICNIRP: 5000 V/m; TCO: 10 V/m; US-Congress/EPA: 10 V/m; BUND: 0.5 V/m; studies on oxidative stress, free radicals, melatonin, childhood leukaemia: 10-20 V/m; nature: < 0.0001 V/m

2 AC MAGNETIC FIELDS (Low Frequency, ELF/VLF)

		No Concern	Slight Concern	Severe Concern	Extreme Concern
Flux density in nanotesla	nT	< 20	20 - 100	100 - 500	> 500
in milligauss	mG	< 0.2	0.2 - 1	1 - 5	> 5

Values apply to frequencies up to and around 50 (60) Hz, higher frequencies and predominant harmonics should be assessed more critically. Line current (50-60 Hz) and traction current (16.7 Hz) are recorded separately.

In the case of intense and frequent temporal fluctuations of the magnetic field, data logging needs to be carried out - especially during nighttime - and for the assessment, the 95th percentile is used.

DIN/VDE: occupational 5 000 000 nT, general 400 000 nT; ACGIH occupational TLV: 200 000 nT; ICNIRP: 100 000 nT; Switzerland 1000 nT; WHO: 300-400 nT "possibly carcinogenic"; TCO: 200 nT; US-Congress/EPA: 200 nT; BioInitiative: 100 nT; BUND: 10 nT; nature: < 0.0002 nT

3 RADIOFREQUENCY RADIATION (High Frequency, Electromagnetic Waves)

		No Concern	Slight Concern	Severe Concern	Extreme Concern
Power density in microwatt per square meter	µW/m²	< 0.1	0.1 - 10	10 - 1000	> 1000

Values apply to single RF sources, e.g. GSM, UMTS, WiMAX, TETRA, Radio, Television, DECT cordless phone technology, WLAN..., and refer to peak measurements. They do not apply to radar signals.

More critical RF sources like pulsed or periodic signals (mobile phone technology, DECT, WLAN, digital broadcasting...) should be assessed more seriously, especially in the higher ranges, and less critical RF sources like non-pulsed and non-periodic signals (FM, short, medium, long wave, analog broadcasting...) should be assessed more generously especially in the lower ranges.

Former Building Biology Evaluation Guidelines for RF radiation / HF electromagnetic waves (SBM-2003): pulsed < 0.1 no, 0.1-5 slight, 5-100 strong, > 100 µW/m² extreme anomaly; non-pulsed < 1 no, 1-50 slight, 50-1000 strong, > 1000 µW/m² extreme anomaly

DIN/VDE: occupational up to 100000000 µW/m², general up to 10000000 µW/m²; ICNIRP: up to 10000000 µW/m²; Salzburg Resolution / Vienna Medical Association: 1000 µW/m²; BioInitiative: 1000 µW/m² outdoor; EU-Parliament STOA: 100 µW/m²; Salzburg: 10 µW/m² outdoor, 1 µW/m² indoor; EEG / immune effects: 1000 µW/m²; sensitivity threshold of mobile phones: < 0.001 µW/m²; nature < 0.000001 µW/m²

Building Biology Evaluation Guidelines for Sleeping Areas SBM-2008, Page 2		No Concern	Slight Concern	Severe Concern	Extreme Concern

4 DC ELECTRIC FIELDS (Electrostatics)

		No Concern	Slight Concern	Severe Concern	Extreme Concern
Surface potential in volt	V	< 100	100 - 500	500 - 2000	> 2000
Discharge time in seconds	s	< 10	10 - 30	30 - 60	> 60

Values apply to prominent materials and appliances close to the body and/or to dominating surfaces at ca. 50 % r.h.

TCO: 500 V; damage of electronic parts: from 100 V; painful shocks and actual sparks: from 2000-3000 V; synthetic materials, plastic finishes: up to 10 000 V; synthetic flooring, laminate: up to 20 000 V; TV screens: up to 30 000 V; nature: < 100 V

5 DC MAGNETIC FIELDS (Magnetostatics)

		No Concern	Slight Concern	Severe Concern	Extreme Concern
Deviation of flux density (steel) in microtesla	µT	< 1	1 - 5	5 - 20	> 20
Fluctuation of flux density (current) in microtesla	µT	< 1	1 - 2	2 - 10	> 10
Deviation of compass needle in degree	°	< 2	2 - 10	10 - 100	> 100

Values for the deviation of the flux density in µT apply to metal/steel and for the fluctuation of the flux density to direct current.

DIN/VDE: occupational 67 900 µT, general 21 200 µT; USA/Austria: 5000-200 000 µT; MRI: 2-4 T; earth's magnetic field: across temperate latitudes 40-50 µT, equator 25 µT, north/south pole 65 µT; eye: 0.0001 nT, brain: 0.001 nT, heart: 0.05 nT; animal navigation: 1 nT; 1 µT = 10 mG

APPENDIX F

Summary of Observed Effects[9]

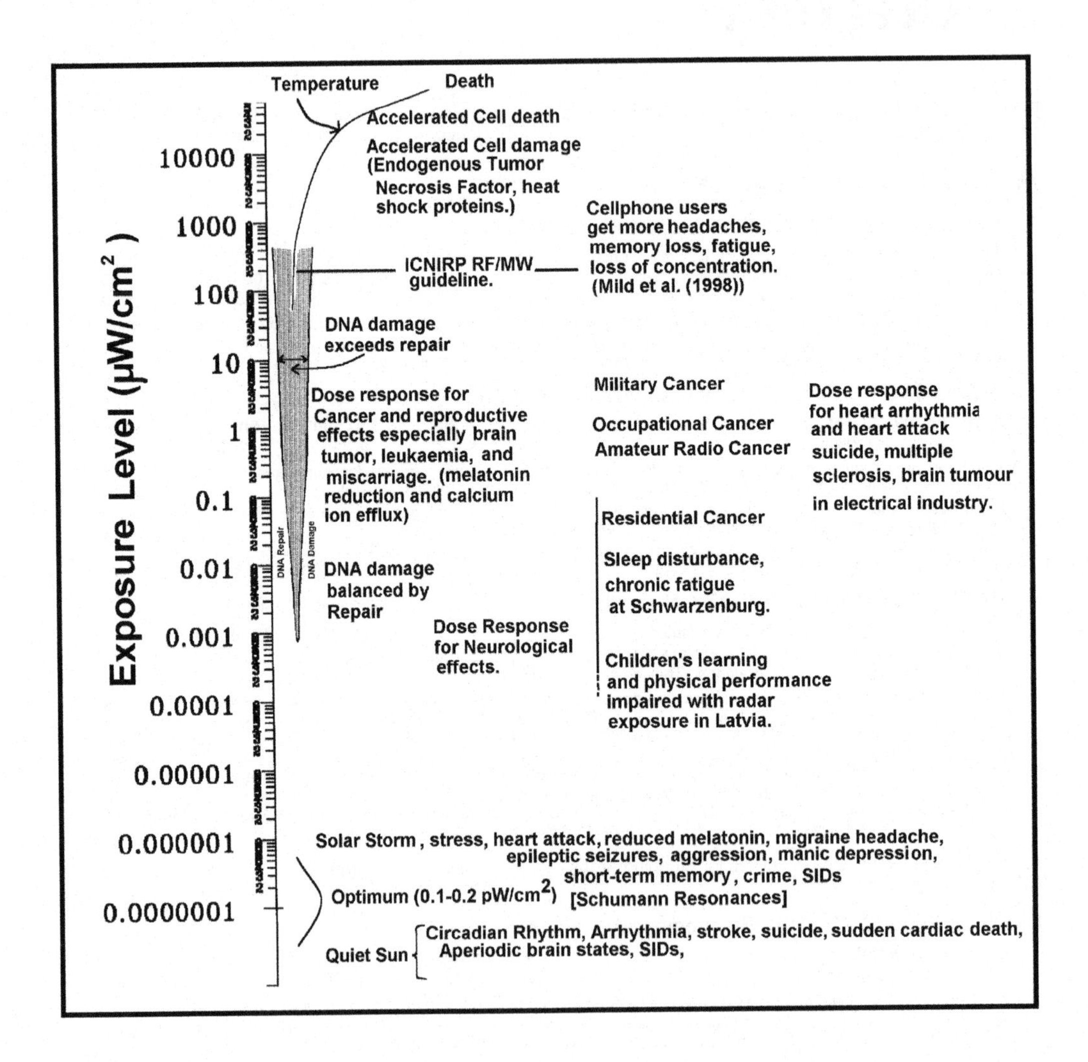

APPENDIX G

Chakras

CHAKRA	NAME	SANSKRIT	LOCATION	COLOUR	ATTRIBUTES
1ST	ROOT	MULHADHARA	RECTUM, TIP OF TAILBONE	RED	GROUNDING, SURVIVAL, BELONGING
2ND	SACRAL	SVADISTHANA	SEVERAL INCHES BELOW THE NAVEL	ORANGE	SEXUALITY, COMPULSION-ADDICTION, CREATIVITY
3RD	SOLAR PLEXUS	MANIPURA	JUST ABOVE THE NAVEL, SOLAR PLEXUS	YELLOW	WILL-POWER, 'GUT-INSTINCT', INNER POWER
4TH	HEART	ANAHATA	CENTRE OF CHEST	GREEN	LOVE, INFINITY
5TH	THROAT	VISHUDDHA	THROAT	BLUE	EXPRESSION
6TH	THIRD EYE	AJNA	THIRD EYE	INDIGO	INTUITION
7TH	CROWN	SAHASRARA	CROWN OF HEAD	VIOLET	DIVINE CONNECTION

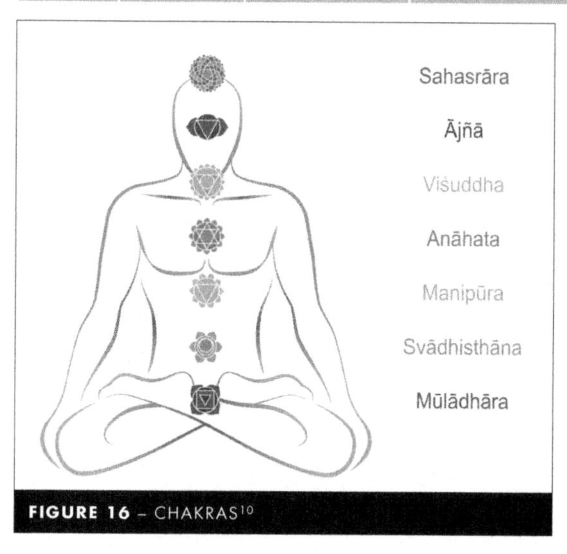

FIGURE 16 – CHAKRAS[10]

APPENDIX H

Radiation Reduction Strategies

MOBILE PHONES

- Pregnant women and children to use landlines to reduce exposure.
- Minimise use to 10 minutes per day – as an emergency device and arms-length texting.
- Switch off in the car.
- Switch off in the home 8 pm to 8 am. When switched on in the home place > 10 m away.
- Deactivate Data Mode and Wi-Fi unless required.
- Quit! Join with others switching off their mobile device as part of Mobile Free Day celebrations – visit mobilefreeday.org.

MOBILE TABLET

- Use on a table if you must use (rather than on the lap and proximal to reproductive organs).
- Discourage use. Ask others to switch off.

WEARABLES

- Don't purchase anything marketed to be worn for extended periods.
- Wearables are a prelude to transhumanism (the next step from 24/7 devices is embedded devices) – help make wearables uncool.
- Share knowledge. Wearables are like strapping a smart phone to our body.
- Request all devices including wearables are switched off in the workplace.

CORDLESS (DECT) PHONES

- Go to corded.

HOME WI-FI

- Switch off when not in use.
- Spend $20 to hardwire.
- Check Telstra wireless is off even if the light indicates it is. Steps in Chapter 11 – Wi-Fi Strong-Armed by the Telstra Mob.

WI-FI IN SCHOOL

- Generate a conversation with other parents, teachers and the principal.
- Observe and ask if children are symptomatic.
- Bring parents and teachers together for an EMF Essentials and Kids talk tailored for school teachers and parents - visit dharamhouse.com.

NBN BROADBAND ANTENNAS

- Ask local council where NBN have a DA for tower construction.
- Rally neighbours, send in submissions and make community noise.
- EMF Essentials talks inform local council thereby aiding quality decision making on behalf of the community – visit dharamhouse.com.

TELSTRA, OPTUS, VODAPHONE, TPG ANTENNAS

- As above but ask council where Telstra, Optus, Vodaphone, TPG have a DA for tower or rooftop antenna construction.

ANTENNAS NEAR RESIDENCE

- Shielding or moving may be required.
- Rally neighbours and send in submissions. We all deserve the right to health.

SMART METERS

- Refuse installation. Do not sign anything. Seek online resources and support.
- Sleep as far from the smart meter as possible.

BABY MONITORS, SMART NAPPIES AND DUMMIES

- Sometimes we are not equipped with knowledge to make empowered healthy choices on behalf of our family. Pass this book to friends swept up by these technologies.

ENERGY CONSERVING LIGHT BULBS

- Carefully remove (mercury is toxic if breakage occurs) and send to toxic waste.
- Halogen globes are a replacement.

POWER LINES

- Fields drop off rapidly with distance. Sleep in areas furthest from the lines.
- Get the wiring diagram for your property. Children's sandpits and play areas need to be moved away from buried power lines.
- Magnetic fields from high voltage lines may be cause for concern.

ELECTRICAL FIELDS

- Wiring inside the wall at the bedhead generates an electric field E. A demand switch installed by an electrician will cut out the circuit when lights are out and electricity is no longer in use. New homes can be strategically wired to eliminate this issue.
- Dirty electricity is associated with dimmer switches, inverters, wireless technology, smart appliances, TVs and mobile chargers. Transients and harmonics typically in the 4 to 100 kHz range can be cleaned up with a series of filters.

COMPUTER

- Hardwire all devices including keyboard and mouse. This also creates distance.
- The hard drive emits ELF-EMF. As with the tablet work from a table rather than lap.

SLEEPING

- Create a Zero EMF Sanctuary in the bedroom. In high exposure scenarios shielding or a Faraday cage 'tent' over the bed may be required.
- Switch computers and phones off at least an hour before bed.
- Harmonise earth energies – visit dharamhouse.com
- Replace spring mattresses with eco-materials such as latex.
- Read THE TRAUMA KEY book and join us to work with your internal held stress and trauma – visit dharamhouse.com.

Notes

INTRODUCTION

1 S Poulter, 'Apple declares biggest profits ever made by a company – amid claims the corporate behemoth isn't paying its share of tax', Daily Mail Australia, 29 January 2015, viewed 30 January 2015, http://www.dailymail.co.uk/news/article-2930588/Apple-declares-biggest-profits-company-amid-claims-corporate-behemoth-isn-t-paying-share-tax.html#ixzz3VAS6q7uj.

PART I – INVISIBLE ENERGIES

1 Australian Associated Press, 'British American Tobacco Profit Soars', *The Australian*, 1 March 2013, viewed 14 November 2014, http://www.theaustralian.com.au/business/latest/british-american-tobacco-profit-soars/story-e6frg90f-1226588047452.

2 Reporter P Taylor, *The Seduction of Smoking: Four Corners Special*, ABC, viewed 29 September 2014 on Australian ABC TV.

3 WHO International Agency for Research on Cancer, 'IARC Classifies Radiofrequency Electromagnetic Fields as Possibly Carcinogenic to Humans', Press Release no. 208, 31 May 2011.

4 Disney, *Disney Fairies Trail*, viewed 15 May 2015, http://www.disney.com.au/disneyfairiestrail/.

5 The Australian Communications and Media Authority, *Regulatory Responsibility*, viewed 15 May 2015, http://www.acma.gov.au/theACMA/About/Corporate/Responsibilities/regulation-responsibilities-acma.

6 The Australian Communications and Media Authority, viewed 17 November 2013, http://acma.gov.au/.

7 C Sage and DO Carpenter – Editors and BioInitiative Working Group, Section 1 Summary for the Public and Conclusions, Summary for the Public, *BioInitiative Report: A Rationale for Biologically-based Public Exposure Standards for Electromagnetic Radiation*, 31 December 2012.

8 S Woodward, 'Trends in cigarette consumption in Australia', *Australia and New Zealand Journal of Medicine*, August 1984, vol. 14, no.4, pp. 405–407.

9 N Gray and D Hill, 'Patterns of tobacco smoking in Australia', *Medical Journal of Australia*, 1975, vol. 2, no. 22, pp. 819–822.

10 Australian Bureau of Statistics, *Gender Indicators, Australia 2011–2012*, cat. No. 4125.0, ABS, Canberra, January 2013.

11 AFP, 'Passive smoking kills 600,000 a year: study', ABC News, 26 November 2010, viewed 20 December 2013, http://www.abc.net.au/news/2010-11-26/passive-smoking-kills-600000-a-year-study/2352570.

12 C Bowen and T Plibersek, 'Government to Increase Tobacco Excise', Australian Government – The Treasury, Media Release # 015, viewed 23 March 2014, http://ministers.treasury.gov.au/DisplayDocs.aspx?doc=pressreleases/2013/015.htm&pageID=003&min=cebb&Year=&DocType=0.

13 Associated Press, 'Altria to spin off Philip Morris International', NBC News, 29 August 2007, viewed 29 November 2013, http://www.nbcnews.com/id/20494757/ns/business-world_business/t/altria-spin-philip-morris-international/#.UsNvWvQW3t-.

Altria Group Inc. Executive profile: Martin J Barrington at fiscal year 2013, Bloomberg Business, viewed 13 February 2015, http://www.bloomberg.com/research/stocks/people/person.asp?personId=30786073&ticker=MO.

14 Productivity Commission Inquiry Report into Gambling, Number 50, 25 February 2010. 'Problem Gambling The Facts', Australian Government, viewed 13 October 2014, http://www.problemgambling.gov.au/facts/.

15 Nick Xenophon, 'Poker Machines', viewed 13 October 2014, http://www.nickxenophon.com.au/campaigns/poker-machines/

16 Ibid.

17 M Chowan, *Quantum Theory Cannot Hurt You*, Faber and Faber, 2007.

18 National Aeronautics and Space Administration, What is the Universe Made Of?, viewed 3 September 2014, http://map.gsfc.nasa.gov/universe/uni_matter.html.

19 M Talbot, *The Holographic Universe*, Harper Collins, 2011, p. 42.

20 S Kahili King, *Earth Energies – A Quest for the Hidden Power of the Planet*, Quest Books, 1992, p. 23.

21 A Wang, *Cosmology and Political Culture in Early China*, Cambridge University Press, 2000.

22 © Commonwealth of Australia 2013 as represented by the Australian Radiation Protection and Nuclear Safety Agency (ARPANSA). http://www.arpansa.gov.au/pubs/emr/spectrum.pdf.

23 R Kane, *Cellular Telephone Russian Roulette: A Historical and Scientific Perspective*, Vantage Press, 2001, p. 47.

SM Michaelson, 'Human Exposure to Nonionizing Radiant Energy— Potential Hazards and Safety Standards', *Proceedings of the IEEE*, April 1972, pp. 389–421.

24 N Steneck, *The Microwave Debate*, MIT Press, 1984, p. 107.

JR Hamer, 'Biological Entrainment of the Human Brain by Low Frequency Radiation', August 1965, obtained by Steneck from ARPA files.

25 N Steneck, p. 108.

RS Cesaro, 'Program Plan No. 562, Pandora', Advanced Sensors Program, ARPA, 15 October 1965.

26 N Steneck, p. 109.

RS Cesaro, 'Initial Test Results', 15 December 1966.

27 N Steneck, p. 109.

RS Cesaro, 'Initial Test Results', 20 December 1966.

28 Credit – Haade, Wikimedia Commons, http://commons.wikimedia.org/wiki/File:Interference_of_two_waves.png.

29 Credit – F Perez, Wikimedia Commons, http://commons.wikimedia.org/wiki/File:Interferences_plane_waves.jpg.

30 © Thinkstock, paid photo reproduced with permission.

31 © Thinkstock, paid photo reproduced with permission.

32 *Oxford English Reference Dictionary*, Revised Second Edition, Edited J Pearsall, B Trumble, Oxford University Press, 2003.

33 O Johansson, Letter on Electrohypersensitivity (EHS), 25 November 2012, viewed 13 December 2013, http://mieuxprevenir.blogspot.com.au/2012/11/dr-olle-johanssons-letter-on.html.

34 R Aidan Martin, 'Biology of Sharks and Rays – Electroreception', Elasmo Research, viewed 18 November 2014, http://www.elasmo-research.org/education/white_shark/electroreception.htm.

35 A Fauteux, 'Electrosensitivity caused by chronic nervous system arousal – Dr Roy Fox', La Maison, 1 May 2012, viewed 20 December 2014, https://maisonsaine.ca/sante-et-securite/electrosmog/ehs-royfo.html.

36 Credit – Peppergrower, Wikimedia Commons, http://commons.wikimedia.org/wiki/File:Phase_shift.svg.

37 Credit – M Reyes and Basecamp, Waveform Illustration, 2012, http://basecamp.com.

38 S Mortavazi et. al., 'Alterations in TSH and Thyroid Hormones following Mobile Phone Use', *Oman Med J.*, October 2009; vol. 24, no. 4, pp. 274–278.

PART II – DEATH TOWERS AND OBSTACLES TO CHANGE

1 C Sage and DO Carpenter, op.cit.

2 A Dullforce, 'FT500 2013', *Financial Times*, 22 June 2013, viewed 29 February 2014, http://www.ft.com/intl/cms/s/0/16f6d1bc-f2c4-11e2-a203-00144feabdc0.html#axzz2qErHSBPj.

3 Credit – B Nowland, An indicative pattern by using antenna design software – numerous variables will alter the actual power.

4 Environmental EME Report 2022002, 1 Newland Street, Bondi Junction, 8 October 2013, viewed 15 April 2014, http://www.rfnsa.com.au/nsa/index.cgi.

5 Environmental EME Report 2022003, 130 Denison Street, Bondi Junction, 20 March 2014, viewed 15 April 2014, http://www.rfnsa.com.au/nsa/index.cgi.

6 C Sage and DO Carpenter, op.cit.

7 AK Datta, *Handbook of Microwave Technology for Food Application*, CRC Press, 2001, p. 216.

8 Litton Industries, 'For Heat, Tune to 915 or 2450 Megacycles', 1965, viewed 13 September 2014, http://www.smecc.org/litton_-_for_heat,_tune_to_915_or_2450_megacycles.htm.

9 JF Burns, 'German Drug Maker Apologizes to Victims of Thalidomide', *NY Times*, 1 September 2012, viewed 12 September 2014, http://www.nytimes.com/2012/09/02/world/europe/grunenthal-group-apologizes-to-thalidomide-victims.html?_r=0.

10 Electromagnetic Hypersensitivity Public Hearing, *European Economic and Social Committee*, Brussels, Belgium, 4 November 2014, viewed 23 December 2014, http://www.eesc.europa.eu/?i=portal.en.events-and-activities-electromagnetic-hypersensitivity.

11 *Electrosensitivity UK News*, vol. 11, no. 2, July 2013, viewed 23 December 2014, http://www.es-uk.info/attachments/article/9/esuknewsJun13b2.pdf.

Austrian Medical Association Guidelines for Diagnosing and Treating Patients with Electrohypersensitivity, Consensus paper of the Austrian Medical Association's EMF Working Group, 3 March 2012.

PART III – LIMITS, LEGALS, LUNACY AND LIES

1 N Chilkov, 'The Link Between Grilled Foods and Cancer', Huffington Post, 22 August 2012, viewed 23 February 2015, http://www.huffingtonpost.com/nalini-chilkov/grilling-health_b_1796567.html.

2 R Kane, p. 43.

HP Schwan and GM Piersol, 'The Absorption of Electromagnetic Energy in Body Tissues', *International Review of Physical Medicine and Rehabilitation*, June 1955, pp. 424–448.

3 R Kane, p. 45.

4 C Sage and DO Carpenter, op. cit.

5 Ibid.

6 ARPANSA, 'Maximum Exposure Levels to Radiofrequency Fields – 3kHz to 300 GHz', ARPANSA Radiation Protection Standard, May 2002, p. 12.

7 Environmental EME Report 2481013, 4 Jonson St, Byron Bay, 26 November 2014, viewed 20 December 2014, http://www.rfnsa.com.au/nsa/index.cgi.

8 Environmental EME Report 2481003, Paterson St, Byron Bay, 7 November 2014, viewed 20 December 2014, http://www.rfnsa.com.au/nsa/index.cgi.

9 C Sage and DO Carpenter, op. cit.

10 PR Log Press Release, 'LTE Cell Phone Radiation Affects Brain Activity in Cell Phone Users', 23 September 2013, viewed 29 December 2013, http://www.prlog.org/12215083-lte-cell-phone-radiation-affects-brain-activity-in-cell-phone-users.html.

11 J Hinks, '5 things you should know about 5G', Tech Radar, 23 March 2015, viewed 25 March 2015, http://www.techradar.com/au/news/world-of-tech/future-tech/5-things-you-should-know-about-5g-1288071.

12 PJ Kiger, 'Fukushima's Radioactive Water Leak: What You Should Know', *National Geographic*, 7 August 2013, viewed 14 March 2014, http://news.nationalgeographic.com.au/news/energy/2013/08/130807-fukushima-radioactive-water-leak/.

13 J Apsley II, 'Radiation Crises and Antidotes', viewed 13 November 2014, http://www.drapsley.com/pages/radiationcrisesantidote.aspx.

14 M Tomonaga, 'Leukaemia in Nagasaki atomic bomb survivors from 1945 through 1959', *Bull World Health Organ.*,

vol. 26, no. 5, 1962, pp. 619–631.

15 P Kaatsch et. al., 'Leukaemia in young children living in the vicinity of German nuclear power plants', *Int. J. Cancer*, vol. 1220, Wiley-Liss, Inc., 2008, pp. 721–726.

16 World Health Organization, 'Electromagnetic fields and public health', June 2007, viewed 10 June 2014, http://www.who.int/peh-emf/publications/facts/fs322/en/.

17 RR Monson (Chair) et. al., 'Health Risks from Exposure to Low Levels of Ionizing Radiation: BEIR VII – Phase 2', The National Academies Press, 2006, p. 340.

18 BioInitiative Report 2012, The Cellular Stress Response: EMF-DNA Interaction – 2012 Supplement, M Blank, p. 2.

19 C Sage and DO Carpenter, op. cit.

20 BioInitiative Report 2012, Section 6 Genetic Effects of Non-Ionizing Electromagnetic Fields – 2012 Supplement, H Lai, p. 2.

21 D Maisch, 'The Procrustean Approach – Setting Exposure Standards for Telecommunications Frequency Electromagnetic Radiation', University of Wollongong, PhD dissertation, 2010, p. 87.

22 N Steneck, pp. 49–50.

23 D Maisch, p. 95.

AA Letavet and ZV Gordon, 'The Biological Action of Ultrahigh Frequencies', *USSR: Academy of Medical Sciences*, 1960. (English edition by the U.S. Joint Publications Research Service.)

24 L Dalton, *Radiation Exposures – The hidden story of the health hazards behind official 'safety' standards*, Scribe Publications, 1991, p. 31.

25 Ibid.

26 J Goldsmith, 'Where the trail leads...Ethical problems arising when the trail of professional work lead to evidence of cover-up of serious risk and mis-representation of scientific judgement concerning human exposures to radar', *Eubios Journal of Asian and International Bioethics*, vol. 5, July 1995, pp. 92-94.

27 D Maisch, p. 194.

A Doull and C Curtain, 'A Case For Reducing Human Exposure Limits Based On Low Level, Non Thermal Biological Effects', unpublished, 1994, p. 1.

28 L Dalton, p.39–41

29 L Dalton, p.42.

DL Hollway, 'The Australian Safety Standard for RF Radiation – A Curate's Egg', *Division of Applied Physics*, CSIRO, 1985.

30 D Maisch, pp. 213–214.

Correspondence with Betty Venables, convenor of the The Electromagnetic Radiation Alliance of Australia (EMRAA) Sutherland Shire Environment Centre, Sutherland, NSW, 27 July 2003.

31 E Rothenberg, 'The Impact of Electromagnetic Fields on Property Prices', Morgan, Lewis and Bockius LLP White Paper, May 1994.

32 'Precedent-Setting Smart Grid Verdict: Jury Awards Vermont Couple $1 mil for Microwave Taking of Their Mountaintop Home, Stop Smart Meters!', 15 December 2013, viewed 18 December 2013, http://stopsmartmeters.org/2013/12/15/precedent-setting-smart-grid-verdict-jury-awards-vermont-couple-1-mil-for-microwave-taking-of-their-mountaintop-home/.

33 Safe EMR blog, 8 August 2014, viewed 2 December 2014, http://www.saferemr.com/2014/08/major-breakthrough-in-cellphone.html.

34 McDonald and Comcare AATA 105, Administrative Appeals Tribunal of Australia, 28 February 2013, viewed 23 December 2013, http://www.austlii.edu.au/au/cases/cth/aat/2013/105.html.

35 Survey by the National Institute for Science, Law & Public Policy Indicates Cell Towers and Antennas Negatively Impact Interest in Real Estate Properties, Business Wire, 3 July 2014, viewed 10 September 2014, http://www.businesswire.com/news/home/20140703005726/en/Survey-National-Institute-Science-Law-Public-Policy#.VP_QFPyUfCs.

PART IV – THE SPIRITUAL EFFECTS OF EMR

1 RK Mishra, 'Electromagnetic Toxins', *LA Yoga Magazine*, November 2010.

2 N Ozaniec, *Chakras – A Beginners Guide*, Hodder and Stoughton, 1999, p. ix.

3 D Childre and H Martin, *The HeartMath Solution*, 2000, Harper San Francisco, p.33.

4 SH Buhner, *The Secret Teachings of Plants: The Intelligence of the Heart in the Direct Perception of Nature*, Bear and Company, 2004, p. 82.

(referencing R McCraty)

5 SH Buhner, p. 84.

6 Ibid., p. 86.

7 Ibid., p. 87.

8 Ibid., p. 90.

9 D Childre and H Martin, op. cit., p.27.

10 Ibid., pp. 28, 29.

11 Ibid., p. 33.

12 © Institute of HeartMath, R McCraty et al, *The Coherent Heart: Heart–Brain Interactions*, Psychophysiological Coherence, and the Emergence of System-Wide Order', 2006, p. 7.

13 VV Hunt interview with S Barber, 'The Promise of Bioenergy Fields and the End to All Disease', *The Spirit of Maat*, vol.2, 2000, viewed 12 January 2014, http://www.spiritofmaat.com/archive/nov1/vh.htm

14 O Johansson (Chair), The Seletun Scientific Statement, 3 February 2011.

15 H Lai, op. cit.

16 C Bird, 'Dowsing in Industry: Hoffman-La-Roche', *The American Dowser*, August 1975.

17 Research on Geopathic Stress, Positive Energy, viewed 12 January 2014, http://www.positiveenergy.ie/research-on-geopathic-stress.

18 S Nicolic, *You Can Heal Yourself: The Definitive Guide to Energy Healing*, Hay House, 2012, p. 149.

DR Cowan and R Girdlestone, *Safe as Houses*, 1995.

19 Cancer Australia, Cancer in Australia Statistics, viewed 20 September 2014, http://canceraustralia.gov.au/affect-

ed-cancer/what-cancer/cancer-australia-statistics.

20 S Nicolic, *You Can Heal Yourself: The Definitive Guide to Energy Healing*, Hay House, 2012, p. 149.

DR Cowan and R Girdlestone, *Safe as Houses*, 1995.

21 K Bachler, Institute for Geopathology SA, viewed 13 January 2014, http://geopathology-za.wikidot.com/kaethe-bachler.

K Bachler, *Earth Radiation*, Wordmasters Ltd., 1989.

22 M Figueiro and MS Rea, 'The Impact of Self-Luminous Displays on Evening Melatonin Levels', *SLEEP*, vol. 36, 2013.

23 L Krahn and IA Gordon, 'In Bed With A Mobile Device: Are The Light Levels Necessarily Too Bright For Sleep Initiation?', *SLEEP*, vol. 36, 2013.

24 Australian Bureau of Statistics, *National Health Survey: Use of Medications, Australia 1995*, cat. No. 4377.0.

25 'Statin Drugs Given for 5 Years for Heart Disease Prevention (Without Known Heart Disease)', *The NNT*, viewed 5 March 2015, http://www.thennt.com/nnt/statins-for-heart-disease-prevention-without-prior-heart-disease/.

26 F Adlkofer, Verum-Foundation Press Release, 6 October 2007, viewed 2 December 2014, http://www.next-up.org/pdf/PressReleaseConcernPrFranzAdelkoferVerumFoundation06102007.pdf

27. N Swaminathan, 'Glia - The Other Brain Cells', *Discover*, Jan – Feb 2011.

28. JJ Rodriguez et. al., 'Astroglia in dementia and Alzheimer's disease', *Cell Death and Differentiation*, vol.16, 2009, pp. 378–385.

29 WHO International Agency for Research on Cancer, 'Interphone study reports on mobile phone use and brain cancer risk', Press Release no. 200, 17 May 2010, http://www.iarc.fr/en/media-centre/pr/2010/pdfs/pr200_E.pdf.

30 N Edgar and E Sibille, 'A putative functional role for oligodendrocytes in mood regulation', *Translational Psychiatry* (2012) 2, e109; doi:10.1038/tp.2012.34, 1 May 2012

31 Elsevier, 'Tripping the Switches on Brain Growth to Treat Depression', Press Release, 15 August 2012, viewed 4 January 2015, http://www.elsevier.com/about/press-releases/research-and-journals/tripping-the-switches-on-brain-growth-to-treat-depression.

32 J Arendt, *Melatonin and the Mammalian Pineal Gland*, Chapman and Hall, 31 December 1994, p. 16.

33 Cottonwood Research Foundation Inc, 'NEW: DMT Found in the Pineal Gland of Live Rats', 23 May 2013, viewed 30 December 2013, http://www.cottonwoodresearch.org/dmt-pineal-2013/.

34 JB Burch et. al., 'Reduced Excretion of a Melatonin Metabolite in Workers Exposed to 60 Hz Magnetic Fields', *Am. J. Epidemiol.*, vol. 150, no. 1, 1999, pp. 27–36.

35 Cancer Council NSW, 'Brain Cancer Action', viewed 20 November 2014, http://braincanceraction.com.au/.

36 WHO International Agency for Research on Cancer, 'Interphone study reports on mobile phone use and brain cancer risk', Press Release no. 200, 17 May 2010, http://www.iarc.fr/en/media-centre/pr/2010/pdfs/pr200_E.pdf.

37 VG Khurana et. al., 'Cell phones and brain tumors: a review including the long-term epidemiologic data', *Surgical Neurology*, vol. 72, no. 3, September 2009, pp. 205-214.

38 WHO International Agency for Research on Cancer, 'IARC Classifies Radiofrequency Electromagnetic Fields as Possibly Carcinogenic to Humans', Press Release no. 208, 31 May 2011.

39 C Teo, 'What if your mobile phone is giving you brain cancer?', *The Punch*, 7 May 2012, viewed 22 December 2013, http://www.thepunch.com.au/articles/what-if-your-mobile-phone-is-giving-you-brain-cancer/.

40 C Althaus, 'Counterpunch: there's no proof mobiles cause cancer', *The Punch*, 8 May 2012, viewed 22 December 2013, http://www.thepunch.com.au/articles/counter-punch-theres-no-proof-mobiles-cause-cancer/.

41 C Smythe, ''Tidal wave' of cancer threatens world', *The Australian*, 4 February 2014, viewed 14 April 2014, http://www.theaustralian.com.au/news/world/tidal-wave-of-cancer-threatens-world/story-fnb64oi6-1226817431313.

42 CA Ross, 'The sham ECT literature: implications for consent to ECT', *Ethical Human Psychiatry and Psychology*, vol. 8, no.1, 2006, pp. 17–28.

43 HA Sackeim et. al., 'The Cognitive Effects of Electroconvulsive Therapy in Community Settings', *Neuropsychopharmacology*, vol. 32, 2007, pp. 244–254.

44 R Ouellet-Hellstrom and WF Stewart, 'Miscarriages among Female Physical Therapists Who Report Using Radio- and Microwave-frequency Electromagnetic Radiation', *Am. J. Epidemiol.*, vol. 138, no. 10, 1993, pp. 775-786.

45 C Bird, 'What has become of the Rife Microscope?', *New Age Journal*, Boston, March 1976, pp. 41–47.

SA Ross, *And Nothing Happened...But You Can Make It Happen!*, Steven A Ross, 2010.

46 G. Lakhovsky, *Radiations and Waves Sources of Our Life*, Emile L Cabella, 1941, p. 35.

47 Ibid., p. 105.

48 Ibid.

PART V – OUR WORLD, OUR CHILDREN

1 O Goldhill, '4G coverage on Mt Everest', *The Telegraph*, 5 July 2013, viewed 15 April 2014, http://www.telegraph.co.uk/technology/mobile-phones/10161553/4G-coverage-on-Mount-Everest.html.

2 Global Partnership on Output-Based Aid, 'Herders Call Home GPOBA pilot connects Mongolia's farmers', *Handshake – International Finance Corporation for the World Bank Group quarterly journal on public-private partnerships*, no. 5, April 2012, p. 62.

3 K Schneider, Mongolia Copper Mine at Oyu Tolgoi Tests Water Supply and Young Democracy, Circle of Blue, 5 November 2013, viewed 22 December 2013, http://www.circleofblue.org/waternews/2013/world/mongolia-copper-mine-oyu-tolgoi-tests-water-supply-young-democracy/.

4 M Ketti, 'Open letter to the IT Minister Anna-Karin Hatt', 18 June 2014, Electrosensitive Association Sweden, viewed 2 November 2014, https://eloverkanslig.org/oppet-brev-till-minister-anna-karin-hatt/.

5 A Fragopoulou et. al., 'Scientific panel on electromagnetic field health risks: consensus points, recommendations, and rationales', *Reviews on Environmental Health*, vol. 25, no. 4, Oct-Dec 2010, pp. 307–317.

6 O Johansson (Chair), The Seletun Scientific Statement, 3

February 2011.

7 Electromagnetic Hypersensitivity Public Hearing, *European Economic and Social Committee*, Brussels, Belgium, 4 November 2014, viewed 23 December 2014, http://www.eesc.europa.eu/?i=portal.en.events-and-activities-electromagnetic-hypersensitivity.

8 B Hernández Bataller (Rapporteur), 'Draft Opinion of the Section for Transport, Energy, Infrastructure and the Information Society on Electromagnetic hypersensitivity', TEN/559, *European Economic and Social Committee*, 19 December 2014.

9 R Stam, 'Comparison of international policies on electromagnetic fields (power frequency and radiofrequency fields)', *National Institute for Public Health and the Environment – Ministry of Health Welfare and Sport*, May 2011.

10 ACEBR, Publications and Conference Extracts, viewed 19 March 2015, http://acebr.uow.edu.au/publications/index.html.

11 NHMRC grants for EME research – grant summaries, Australian Government, viewed 19 March 2015, http://www.nhmrc.gov.au/grants-funding/outcomes-funding-rounds/nhmrc-funded-research-effects-electromagnetic-energy/nhmrc-gr.

12 Wireless LANs, CSIRO, 5 May 2009 updated 1 April 2012, viewed 3 September 2014, http://www.csiro.au/Outcomes/ICT-and-Services/People-and-businesses/wireless-LANs.aspx.

13 D Pauli, 'NSW gets world's largest Wi-Fi network – Government to connect every school in the state', *Computerworld*, 16 March 2010, viewed 12 February 2014, http://www.computerworld.com.au/article/339733/nsw_gets_world_largest_Wi-Fi_network/

14 'PISA report finds Australian teenagers education worse than 10 years ago', *news.com.au*, 4 December 2013, http://www.news.com.au/national/pisa-report-finds-australian-teenagers-education-worse-than-10-years-ago/story-fncynjr2-1226774541525

15 'Dr Marie-Therese Gibson resigns from Tangara School for Girls over Wi-Fi health worries', news.com.au, 29 September 2013, viewed 4 December 2013, http://www.news.com.au/technology/dr-marietherese-gibson-resigns-from-tangara-school-for-girls-over-wifi-health-worries/story-e6frfrnr-1226729172333

16 A Sasco, 'France Passes New National Law!', *Environmental Health Trust*, viewed 13 February 2015, http://ehtrust.org/france-new-national-law-bans-wifi-nursery-school/.

17 'Dr Marie-Therese Gibson resigns from Tangara School for Girls over Wi-Fi health worries', news.com.au, 29 September 2013, viewed 4 December 2013, http://www.news.com.au/technology/dr-marietherese-gibson-resigns-from-tangara-school-for-girls-over-wifi-health-worries/story-e6frfrnr-1226729172333

18 J Huss (Rapporteur) and Committee on the Environment, Agriculture and Local and Regional Affairs, 'The potential dangers of electromagnetic fields and their effect on the environment, Council of Europe Parliamentary Committee, Doc. 12608, 6 May 2011, viewed 29 September 2014, http://www.assembly.coe.int/ASP/Doc/XrefViewPDF.asp?FileID=13137&Language=EN.

19 *Oxford English Reference Dictionary Revised Second Edition*, Edited J. Pearsall, B. Trumble, Oxford University Press, 2003.

20 W Löscher and G Käs (translated by R Riedlinger), 'Conspicuous behavioural abnormalities in a dairy cow herd near a TV and Radio transmitting antenna', *Prakt. Tierarzt*, **vol.** 79, no. 5, 1998, pp. 437-444.

21 A Balmori, 'Mobile Phone Mast Effects on Common Frog (Rana temporaria) Tadpoles: The City Turned into a Laboratory', *Electromagnetic Biology and Medicine*, vol. 29, 2010, p. 31–35.

22 LE Foley et. al., 'Human cryptochrome exhibits light-dependent magnetosensitivity', *Nat. Commun.*, vol. 2, p. 356. 2:356 doi: 10.1038/ncomms1364 (2011).

23 JL Kirschvink et. al., 'Magnetite biomineralization in the human brain', *Proc. Natl. Acad. Sci. USA*, vol. 89, August 1992, pp. 7683-7687.

24 RR Baker et. al., 'Magnetic bones in human sinuses', *Nature*, 1983 Jan 6;301(5895):79–80.

25 FiLIP, viewed 16 May 2015, http://webshop.myfilip.com/.

26 FiLIP, viewed 16 May 2015, http://www.myfilip.com/about-filip/.

27 Ibid.

28 FiLIP, viewed 16 May 2015, http://www.myfilip.com/filip-technology/.

29 'Japan sensor will let nappy say baby needs changing', news.com.au, 10 February 2014, viewed 13 March 2014, http://www.news.com.au/lifestyle/parenting/japan-sensor-will-let-nappy-say-baby-needs-changing/story-fnet085v-1226823074104?from=public_rss.

30 'The world's smartest dummy', Blue Maestro, viewed 14 March 2015, http://bluemaestro.com/pacifi-smart-pacifier/.

31 ARPANSA, 'Mobile Phones and Children', Fact Sheet #11, viewed 14 March 2015, http://www.arpansa.gov.au/pubs/eme/fact11.pdf.

32 OP Gandhi et. al., 'Exposure Limits: The underestimation of absorbed cell phone radiation, especially in children', *Electromagnetic Biology and Medicine*, vol. 31, no. 1, March 2012, pp. 34–51.

33 OP Gandhi, G Lazzi and CM Furse, 'Electromagnetic Absorption in the Human Head and Neck for Mobile Telephones at 835 and 1900 MHz', *IEEE Transactions on Microwave Theory and Techniques*, vol. 44, October 1996, pp. 1884–1897.

34 Reproduced from OP Gandhi, G Lazzi and CM Furse, 'Electromagnetic Absorption in the Human Head and Neck for Mobile Telephones at 835 and 1900 MHz', *IEEE Transactions on Microwave Theory and Techniques*, vol. 44, October 1996, pp. 1884–1897.

PART VI – WEAPONS OF MASS DESTRUCTION

1 A Hoh, 'Phone use overtakes not wearing seatbelts as cause of fatal car accidents', *Sydney Morning Herald*, 13 February 2014, viewed 13 February 2014, http://smh.drive.com.au/motor-news/phone-use-overtakes-not-wearing-seatbelts-as-cause-of-fatal-car-accidents-20140213-32n27.html.

2 S McEvoy et. al., 'Role of mobile phones in motor vehicle crashes resulting in hospital attendance: a case-crossover study', *BMJ* 2005;331:428, 18 August 2005.

3 Lieutenant Colonel TL Thomas, 'The Mind Has No Firewall', *Parameters*, Spring 1998, pp. 84–92.

D Pasternak, 'Wonder Weapons', *U.S. News and World Report*, 7 July 1997, pp. 38–46.

4 Ibid.

5 Australian Institute of Health and Welfare, 'Dementia', viewed 10 October 2014, http://www.aihw.gov.au/dementia/.

6 'Brain diseases affecting more people and starting earlier than ever before', *Science Daily*, 10 May 2013, viewed 10 October 2014, http://www.sciencedaily.com/releases/2013/05/130510075502.htm.

7 15.5 Sponsorship, Quit Victoria, 1995, viewed 17 February 2014, http://www.tobaccoinaustralia.org.au/fandi/fandi/c15s5.htm.

8 'Samsung is world's biggest advertiser spending $US4.3b on ads', *Financial Review*, 29 November 2013, viewed 19 December 2013, http://www.afr.com/p/business/marketing_media/samsung_is_world_biggest_advertiser_LHFpigt3BzcNqDXfA6flQP.

9 B Nowland design. L Swales graphics.

10 S Chapman, 'Big Tobacco crashes at first legal hurdle on packaging', *The Conversation*, 15 August 2012, viewed on 23 December 2013, http://theconversation.com/big-tobacco-crashes-at-first-legal-hurdle-on-plain-packaging-8807.

11 Australian Associated Press, 'Plain packaging prompts quit attempts', *The Australian*, 13 January 2013, viewed on 23 December 2013, http://www.theaustralian.com.au/news/health-science/plain-packaging-prompts-quit-attempts/story-e6frg8y6-1226800573253

12 A Lutz et. al., 'Long-term meditators self-induce high-amplitude gamma synchrony during mental practice', *Proceedings of the National Academy of Sciences*, vol. 101 no. 46, 2004.

13 G Thomas, *Journey Into Madness – The True Story of Secret CIA Mind Control and Medical Abuse*, Bantam Books, 1989, p. 250.

14 JM Delgado, *Physical Control of the Mind: Towards a Psychocivilized Society*, Harper Collins, 28 October 1969.

15 A Frey, 'Human Auditory Response to Modulated Electromagnetic Energy', *J. Appl. Physiol.*, vol. 17, no. 4, pp. 689–692, 1962.

16 ARPANSA, 'Human auditory perception resulting from exposure to high power pulse or modulated microwave radiation— specification of appropriate safety limits', viewed 10 January 2015, http://www.arpansa.gov.au/pubs/rps/aud_perc.pdf

17 Ibid.

18 ARPANSA, 'Maximum Exposure Levels to Radiofrequency Fields – 3kHz to 300 GHz', ARPANSA Radiation Protection Standard, May 2002, p. 8.

19 Rabichev et. al., United States Patent # 3773049 – Apparatus for the Treatment of Neuropsychic and Somatic Diseases with Heat, Light, Sound and VHF Electromagnetic Radiation, 20 November 1973.

20 R Adey and SM Bawin, 'Binding and release of brain calcium by low-level electromagnetic fields: A review', *W. Radio Sci.*, vol. 17, no. 5S, pp. 149S–157S, 1982.

21 US Department of the Army, 'Bioeffects of Nonlethal Weapons', Unclassified 6 December 2006, p. 1.

22 E Bernays, *Propaganda*, Ig Publishing, 1928, pp. 9–10.

23 Australian Mobile Telecommunications Association, viewed 13 February 2015, http://www.amta.org.au.

PART VII – RECOVERY

1 BJ Carducci, *The Psychology of Personality: Viewpoints, Research, and Applications*, 2nd Ed, Wiley Blackwell, 2009, p. 141.

2 Yoda Quotes, ThinkExist.com, viewed 25 November 2014, http://thinkexist.com/quotation/fear_is_the_path_to_the_dark_side-fear_leads_to/255552.html.

3 Lock the Gate Code of Conduct for Protesters, Gasfield Free Byron Shire, viewed 30 November 2014, http://www.gasfieldfreebyronshire.org/code-of-conduct/.

4 C Borland and GS Landrith III, 'Improved quality of life through the Transcendental Meditation program: Decreased crime rate', *Scientific research on the Transcendental Meditation program: Collected papers*, vol. 1. Rheinweiler, Germany, Maharishi European Research University Press, 1977.

5 DW Orme-Johnson et. al., 'Time series impact assessment analysis of reduced international conflict and terrorism: Effects of large assemblies of participants in the Transcendental Meditation and TM-Sidhi program', Presented at the 85th Annual Meeting of the American Political Science Association, Atlanta, Georgia, 1989.

APPENDICES

1 C Sage and DO Carpenter, op. cit.

2 Ibid.

3 O Hallberg and G Oberfeld, 'Letter to the Editor: Will We All Become Electrosensitive?', *Electromagnetic Biology and Medicine*, vol. 25, 2006, pp. 189–191.

4 © Elsevier Science from R Kitchen, RF and Microwave Radiation Safety Handbook, Elsevier Science, September 2001, p.12

5 Ibid., p.13.

6 Credit – G Richards, Wikimedia Commons, http://commons.wikimedia.org/wiki/File:Harmonic_partials_on_strings.svg.

7 Credit – Borb, Wikimedia Commons, http://commons.wikimedia.org/wiki/File:Inverse_square_law.svg.

8 © Baubologie Maes / Institut für Baubiologie + Ökologie IBN (translated by K Gustavs), 'Building Biology Evaluation Guidelines for Sleeping Areas', SBM-2008 7th Edition, 2008, www.baubiologie.de.

9 © N Cherry and Lincoln University, N Cherry, 'Evidence that Electromagnetic Radiation is Genotoxic: The implications for the epidemiology of cancer and cardiac, neurological and reproductive effects' [Extended from a paper to the conference on possible health effects on health of radiofrequency electromagnetic fields European Parliament Brussels], Lincoln University, New Zealand, 2 March 2001.

10 © Thinkstock, paid photo reproduced with permission.

Index

About the Author

BENJAMIN NOWLAND (Honours Mechanical Engineer, Grad. Cert. Environmental Management, Cert. IV Training and Assessment, Certified Health Practitioner and Yoga Teacher) shares unique perspectives on health and spirituality. He has spent two decades exploring human potential and in his writing conveys discovered truths with rawness and simplicity. His company Dharam House (dharamhouse. com) empowers individuals through talks, retreats, books and consultations. Ben's early material success segued into a cataclysmic 12-month period during which he lost everything including his health, wealth and relationships. Profound healing insights were received. It is his mission to deliver this knowledge to others to support their journey.

CPSIA information can be obtained at www.ICGtesting.com
Printed in the USA
BVOW11s1736031215

428894BV00005B/19/P